Advances in Industrial Control

T0140613

Pedro Castillo, Rogelio Lozano and Alejandro E. Dzul

Modelling and Control of Mini-Flying Machines

With 126 Figures

 Springer

Pedro Castillo, PhD
Rogelio Lozano, PhD
Alejandro E. Dzul, PhD

Heuristique et Diagnostic des Système Complexes, UMR-CNRS 6599,
Université de Technologie de Compiègne, Centre de Recherche de Royalieu,
BP 20529, 60200 Compiègne, France

Whilst we have made considerable efforts to contact all holders of copyright material contained in this book, we may have failed to locate some of them. Should holders wish to contact the Publisher, we will be happy to come to some arrangement with them.

British Library Cataloguing in Publication Data
Castillo, Pedro
  Modelling and control of mini-flying machines. – (Advances in industrial control)
  1. Drone aircraft – Control systems 2. Drone aircraft – Control systems – Mathematical models
  3. Airplanes – Automatic control 4. Airplanes – Automatic control – Mathematical models
  I. Title II. Lozano, R. (Rogelio), 1954– III. Dzul, Alejandro E.
  629.1′326

Advances in Industrial Control series ISSN 1430-9491
ISBN 978-1-84996-977-2   e-ISBN 978-1-84628-179-2
Springer Science+Business Media
springeronline.com

Printed in the United States of America
69/3830-543210 Printed on acid-free paper

Advances in Industrial Control

Series Editors

Professor Michael J. Grimble, Professor Emeritus of Industrial Systems and Director
Professor Michael A. Johnson, Professor of Control Systems and Deputy Director

Industrial Control Centre
Department of Electronic and Electrical Engineering
University of Strathclyde
Graham Hills Building
50 George Street
Glasgow G1 1QE
United Kingdom

Series Advisory Board

Professor E.F. Camacho
Escuela Superior de Ingenieros
Universidad de Sevilla
Camino de los Descobrimientos s/n
41092 Sevilla
Spain

Professor S. Engell
Lehrstuhl für Anlagensteuerungstechnik
Fachbereich Chemietechnik
Universität Dortmund
44221 Dortmund
Germany

Professor G. Goodwin
Department of Electrical and Computer Engineering
The University of Newcastle
Callaghan
NSW 2308
Australia

Professor T.J. Harris
Department of Chemical Engineering
Queen's University
Kingston, Ontario
K7L 3N6
Canada

Professor T.H. Lee
Department of Electrical Engineering
National University of Singapore
4 Engineering Drive 3
Singapore 117576

Professor Emeritus O.P. Malik
Department of Electrical and Computer Engineering
University of Calgary
2500, University Drive, NW
Calgary
Alberta
T2N 1N4
Canada

Professor K.-F. Man
Electronic Engineering Department
City University of Hong Kong
Tat Chee Avenue
Kowloon
Hong Kong

Professor G. Olsson
Department of Industrial Electrical Engineering and Automation
Lund Institute of Technology
Box 118
S-221 00 Lund
Sweden

Professor A. Ray
Pennsylvania State University
Department of Mechanical Engineering
0329 Reber Building
University Park
PA 16802
USA

Professor D.E. Seborg
Chemical Engineering
3335 Engineering II
University of California Santa Barbara
Santa Barbara
CA 93106
USA

Doctor I. Yamamoto
Technical Headquarters
Nagasaki Research & Development Center
Mitsubishi Heavy Industries Ltd
5-717-1, Fukahori-Machi
Nagasaki 851-0392
Japan

Foreword

The series Advances in Industrial Control aims to report and encourage technology transfer in control engineering. The rapid development of control technology impacts all areas of the control discipline. New theory, new controllers, actuators, sensors, new industrial processes, computer methods, new applications, new philosophies,..., new challenges. Much of this development work resides in industrial reports, feasibility study papers and the reports of advanced collaborative projects. The series offers an opportunity for researchers to present an extended exposition of such new work in all aspects of industrial control for wider and rapid dissemination.

Nonlinear control theory and applications are the main focus of much recent research within the control community at present. A common sentiment is that to make further performance gains it is necessary to formulate and work with the full nonlinear modelled behaviour of practical systems. Aeronautic applications are one field where nonlinear models are often available and where there is a key necessity to use high performance control systems. However, whilst many control researchers have access to fairly sophisticated nonlinear aeronautical system models, few will have access to working and experimental systems in the field. One solution to this might be to explore the control of mini-flying machines. This Advances in Industrial Control monograph by P. Castillo, R. Lozano and A. Dzul provides just the right sort of inspiration to start such an initiative. It is a fascinating mix of mini-flying machine history, modelling and experimental results from constructed mini flying machines.

The volume opens with a historical review featuring the inventions of Leonardo da Vinci, the success of Heinrich Focke and Gert Achgelis (1936) and the first certified commercial helicopter of Igor Sikorsky (ca1950) supported by some fascinating pictures of these technological developments. Throughout the volume, a number of aeronautical control problems are investigated theoretically and then pursued using various type of mini flying machine experimental platforms. One of these platforms is a quad-rotor rotorcraft used in Chapters 2 and 3 for which experiment videos are available on the web at http://www.hds.utc.fr/~castillo/4r_fr.html.

This rotorcraft is also used in experiments for Chapter 4. Chapter 6 sees the use of a different experimental set-up, a vertical flying stand, which is used to investigate the altitude control of mini- helicopters.

Chapter 7 reviews and investigates the control of the tail-sitter unmanned aerial vehicle; again there is an interesting description of an experimental platform devised by Dr R. H. Stone of the University of Sydney, Australia. The chapter includes the derivation of a six-degrees of freedom model and experimental results.

To compete the set of flying technologies and configurations, autonomous airships are introduced in Chapter 8. This chapter is contributed by Dr Y. Bestaoui at the University of Evry, France and presents modelling and control system results motivated by the experimental platform, the AS200 airship devised by the Laboratoire des Systemes Complexes.

The volume concludes with a very detailed chapter on sensors, modems and micro controllers for unmanned aerial vehicles. These devices are very important for the accurate control of mini-flying machines and the detailed descriptions provide a useful introduction for the novice reader and excellent reference material for the expert.

The industrial, commercial and scientific uses of mini-flying machines are still under development and the aeronautical control engineer interested in this field will find this monograph a valuable source of new modelling and control material. In the academic world, educators are always seeking new and exciting ways of demonstrating control system techniques and technologies; some of the experimental platforms described in this volume seem tailor made for such a role. A most satisfying aspect of this new Advances in Industrial Control monograph is the way aeronautical technology, models, control systems theory and control engineering come together in the experimental results presented and as such, the volume should also appeal to the wide range of control engineering readership.

M.J. Grimble and M.A. Johnson
Industrial Control Centre
Glasgow, Scotland, U.K.
Winter, 2004

Preface

The opportunities to apply control principles and methods to mini-aircraft are flourishing in the beginning of the 21st century. Nonlinear modelling and modern nonlinear control theory play an important role in achieving high performance autonomous flight for new mini- and micro-flying machines. The rapid development is also due to the advances in computation, communication and sensing which are becoming increasingly inexpensive and omnipresent. This will make possible the development of new small UAVs (Unmanned Autonomous Vehicles) with sophisticated aerodynamical configurations. The use of automatic control theory will permit mini-aircraft to be more efficient and cost effective with a degree of intelligence and reactivity that will dramatically expand their field of applications.

This book presents a systematic study of modelling and nonlinear control of aerial vehicles taking into account mini-aircraft physical nonlinearities and aerodynamical forces, sensors and actuators limitations, and the effects of delays due to the computation time. The objective is to obtain control laws which perform satisfactorily even in presence of disturbances commonly encountered in real applications. This book presents also the development of platforms of aerial vehicles and an overview of the main sensors currently used in mini-UAVs.

This book particularly focus on aerial vehicles capable of hovering as helicopters as well as performing forward flight as normal planes. We study the following flying machines:

- A planar vertical take-off and landing vehicle (PVTOL aircraft).
- A four-rotor rotorcraft.[1]
- A reduced model of a classical helicopter.
- A mini-airplane (T-Wing).
- A mini-blimp.

---

[1] Also named quad-rotor rotorcraft.

Chapter 1 presents a brief history of aeronautics and particularly focuses on the evolution of multi-rotor rotorcraft. The success of aerial vehicles today would certainly not be possible without the relentless attempts realized in the last century by many courageous pioneers of the aerospace domain.

The model of the planar vertical take-off and landing aircraft represents a simplification of the model of a real vertical take-off aircraft. It also represents the longitudinal model of a helicopter. The control community has shown great interest in the PVTOL problem because it is an interesting and challenging theoretical problem of nonlinear systems which is clearly motivated by an application. Designing appropriate nonlinear controllers based on the PVTOL model can be used to improve the performance and the stability margins of the closed-loop system for vertical take-off and landing aircraft.

Chapter 2 presents a nonlinear control strategy for stabilizing the PVTOL using the approach of nested saturations. We have developed an experimental platform for testing the control algorithms we have proposed.

The four-rotor rotorcraft can be viewed as a generalization of the PVTOL in three-dimensional space. It can also be viewed as an alternative to standard helicopters having no swashplate. Four-rotor rotorcraft have fewer mechanical parts than standard mini-helicopters. This results in reduced time spent in maintenance. Given that the blades turn in opposite directions, the gyroscopic phenomena due to the blades is less important in quad-rotor rotorcraft than it is for standard helicopters. As a consequence quad-rotor rotorcraft have larger manoeuvrability but this also means that they tend to be more unstable. In Chapter 3 we present a nonlinear control strategy for stabilizing the helicopter having four rotors. The proposed control law has been tested in real-time experiments.

We have also carried out experiments to test the robustness of the controller with respect to delays which appear in the computation of the control law as well as in the computation of the position and orientation of the aircraft. We have studied in particular the controller used for the yaw angular displacement. We have designed a controller based on the prediction of the state and tested it in real-time flight. It has been shown that the system response remained basically unaffected in spite of significant delay. Chapter 4 presents these results.

In Chapter 5 we present the nonlinear model for the classical helicopter obtained using the Euler–Lagrange technique and also using the Newton laws. We have used the model we have obtained for designing a nonlinear control law based on the backstepping technique.

The control of a helicopter in 3D is an experiment which involves certain risks for the persons performing the study as well as for the equipment. In order to minimize the risks of accident we have built a vertical flying stand that allows the helicopter to move freely vertically and also to turn around the vertical axis. Although the behaviour of the helicopter in such a platform is different to its behaviour in free flight, this platform allows to gain a lot of insight on the problem and to perform experiments with limited risks.

The model of the helicopter during free flight reduces to a simpler model when it operates in the vertical flying stand. The model of the helicopter in the platform can also be obtained directly by using the Euler–Lagrange approach. Chapter 6 presents the model of the helicopter mounted on the vertical flying stand and a control law to stabilize the helicopter altitude and the yaw angular displacement.

For many years researchers have devoted much effort to developing planes that can take-off and land vertically. Such aircraft would require very inexpensive airport facilities. Chapter 7 presents the "T-wing" aircraft which is a new tailsitter plane developed by H. Stone at the University of Sydney. The nonlinear model structure, the identification of the aircraft parameters and the control algorithm for the T-wing are presented. We are indebted to Dr. H. Stone for writing this chapter.

Lighter than air vehicles, or blimps, are suitable for a wide range of applications such as advertising, aerial photography and aerial inspection platforms. They play an important role in environmental and climatologic research which require the development of autonomous airships. Chapter 8 presents a model of a blimp and the application of modern control theory of underactuated nonlinear systems to stabilize the airship. We thank Dr. Y. Bestaoui for writing this chapter.

Sensors are an essential part of the closed-loop control system. Sensors used in UAVs are rapidly changing. New sensors are more precise, lighter and less expensive than ever before. Chapter 9 presents an overview of the sensors that are currently used in mini-aircraft.

We wish to thank the University of Technology of Compiègne, Laboratory Heudiasyc, CNRS, Ministry of Education, Picardie Region in France, CONACYT in México and the bilateral laboratory LAFMAA. The research in UAVs and the experimental platforms that led to this book would not have been possible without their valuable support.

Compiègne, France
July 2004

P. Castillo
R. Lozano
A. Dzul

Contents

1

Introduction and Historical Background

Automatic flying of intelligent vehicles moving in space represents a huge field of applications. We are particularly interested in this book in small vehicles, such as helicopters in all configurations (configuration with a main rotor with or without tail rotor, configuration in tandem, configuration with coaxial opposite rotary rotors, configuration with four, two or one rotors) for their adaptability and manoeuvrability. Helicopters are relatively complex and difficult to control but they allow to carry out various tasks. Automating these machines allows among other things to guarantee a minimum of security when the pilot is no longer able to control the vehicle. Automatic flying can also be used when the task to be achieved is too repetitive or too difficult for the pilot [122].

Automatic control of small flying machines opens up applications in the fields of security (supervision of aerial space, urban traffic), management of natural risks (supervision of active volcanoes), of environment (measuring air pollution, supervision of forests), for intervention in hostile environments (radioactive atmospheres, removal of mines without human intervention), management of ground installations (dams, lines with high tension, pipelines), agriculture (detection and treatment of infested cultivations), and aerial shooting in the production of movies, to cite a few examples [114, 157].

In this chapter we briefly present the evolution of flying machines that are able to take-off vertically. A historical review is given for various multi-rotor rotorcraft as well as some definitions.

1.1 Definitions

AIRCRAFT

An aircraft is any machine capable of flight. Aircraft can be divided into two categories [185]:

- Heavier: Autogyros, helicopters and variants, and conventional fixed-wing aircraft.

- Lighter: Balloons and airships. The distinction between a balloon and an airship is that an airship has some means of controlling forward motion and steering while balloons simply drift with the wind.

The abbreviation VTOL is applied to aircraft other than helicopters that can take-off or land vertically. Similarly, STOL stands for Short Take Off and Landing.

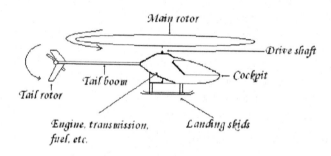

Fig. 1.1. Components of a helicopter.

HELICOPTER

A helicopter is an aircraft that can take-off and land vertically. Also called a *rotary aircraft*, this aircraft can hover and rotate in the air and can move sideways and backwards while aloft. This aerial vehicle can change direction very quickly and can stop moving completely and begin hovering [116].

The helicopter began as a basic principle of rotary-wing aviation. The precision of parts due to the Industrial Revolution enabled the helicopter to evolve into the modern machines we see flying today. The need for accurate machinery and fixtures was evident when the earliest helicopter models lacked the efficiency and flying capability of modern helicopters. The main components of a helicopter are given in Figure 1.1.

UAV

An *Unmanned Aerial Vehicle* (UAV), also called a *drone*, is a self-descriptive term used to describe military and civil applications of the latest generations of pilotless aircraft [185]. UAVs are defined as aircraft without the onboard presence of pilots [179], used to perform intelligence, surveillance, and reconnaissance missions. The technological promise of UAVs is to serve across the full range of missions. UAVs have several basic advantages over manned systems including increased manoeuvrability, reduced cost, reduced radar signatures, longer endurance, and less risk to crews.

1.2 Early Concepts of VTOL Aircraft

"The idea of a vehicle that could lift itself vertically from the ground and hover motionless in the air was probably born at the same time that man first dreamed of flying."[1]

During the past sixty years since their first successful flights, helicopters have matured from unstable, vibrating contraptions that could barely lift the pilot off the ground, into sophisticated machines of quite extraordinary flying capability.

The idea of vertical flight aircraft can be traced back to early Chinese tops, a toy first used about 400 BC. The earliest versions of the Chinese top consisted of feathers at the end of a stick, which was rapidly spun between the hands to generate lift and then released into free flight (see Figure 1.2). These toys were probably inspired by observations of the seeds of trees such as the sycamore, whose whirling, autorotating seeds can be seen to carry on the breeze [47, 61, 64].

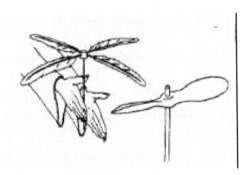

Fig. 1.2. The first concept of rotary-wing aviation.

---

[1] Igor Ivanovitch Sikorsky

"Trovo, se questo strumento a vite sar ben fatto, cio fatto di tela lina, stopata i suoi pori con amido, e svoltata con prestezza, che detta vite si fa la femmina nellaria e monter in alto."[2] "I believe that if this screw device is well manufactured, that is, if it is made of linen cloth, the pores of which have been closed with starch, and if the device is promptly reversed, the screw will engage its gear when in the air and it will rise up on high" [120].

In 1483 Leonardo Da Vinci designed a sophisticated aircraft capable of hovering. Some experts have identified this aircraft as the ancestor of the helicopter. The aircraft called *aerial screw* or *air gyroscope* had a diameter of 5 m (see Figure 1.3), and was operated presumably by four men who might have stood on the central platform and exerted pressure on the bars in front of them with their hands, so as to make the shaft turn.

The main idea was that if an adequate driving force were applied, the machine might have spun in the air and risen off the ground.

Fig. 1.3. The air screw. Credits – Hiller Aviation Museum [66].

In 1754, Mikhail Lomonosov developed a small coaxial rotor similar to the *Chinese top* but powered by a wound-up spring device. The aircraft flew freely and climbed to a good altitude [61, 64].

In 1783, Launoy and Bienvenu used a coaxial version of the Chinese top in a model consisting of a counter-rotating set of turkey feathers [61, 64].

---

[2] Leonardo Da Vinci

A large number of minor inventions contributed to the advancement of the helicopter. Between the fifteenth and twentieth centuries, it was not yet possible to produce the machinery needed to build helicopters, like turbine engines and rotors, but as the Industrial Revolution created factories and technology accelerated, the helicopter evolved.

One of the first breakthroughs in helicopter advancement was by George Cayley who produced a converti-plane in 1843 [146]. Cayley designed an aircraft capable of hovering that he called the "*aerial carriage*" (see Figure 1.4). However, Cayley's device remained an idea because the only powerplants available at the time were steam engines, and these were much too heavy to allow for successful powered flight [61, 64].

Fig. 1.4. Aerial carriage. Credits – Hiller Aviation Museum [66].

The lack of a suitable powerplant continued to stifle aeronautical progress, both for fixed and rotating wing applications, but the use of miniature lightweight steam engines met with some limited success.

In the 1840s, Horatio Phillips constructed a steam-driven vertical flight machine where steam generated by a miniature boiler was ejected out of the blade tips [61, 64].

In the 1860s, Ponton d'Amecourt flew a number of small steam-powered helicopter models (Figure 1.5). He called his machines *helicopteres*, which is a word derived from the Greek adjective *elikoeioas* meaning spiral or winding, and the noun *pteron* meaning feather or wing [61, 64].

Fig. 1.5. Gustave Ponton d'Amecourt's helicopters. Credits – Hiller Aviation Museum [66].

At the beginning of the twentieth century nearly all prior attempts at vertical flight could be considered as inventive, the inherent aerodynamic and mechanical complexities of building a vertical flight aircraft were to challenge many ambitious efforts. A contributing factor was the relatively few scientific investigations of flight or studies into the science of aerodynamics. The history of flight documents literally hundreds of failed helicopter inventions, which either had inadequate installed power or limited control capability, or more often than not, the machine just vibrated itself to pieces [61].

In the 1880s, Thomas Alva Edison experimented with small helicopter models in the United States. He tested several rotor configurations driven by a gun cotton engine, which was an early form of internal combustion engine. However, a series of explosions deterred further efforts with these engines. Later, Edison used an electric motor for power, and he was one of the first to realize from his experiments the need for a large diameter rotor with low blade area to give good hovering efficiency [61, 64].

While one can draw several parallels in the technical development of the helicopter when compared with fixed-wing aircraft, the longer and more tumultuous gestation of vertical flight aircraft is a result of the greater depth of knowledge required before all the various aerodynamic and mechanical problems could be understood and overcome.

By 1920, gasoline powered piston engines with higher power-to-weight ratios were more widely available, and the control problems of achieving successful vertical flight were at the forefront. This era is marked by the development of a vast number of prototype helicopters throughout the world. Most of these machines made short hops into the air or flew slowly forward in ground effect. Many of the early designs were built in Great Britain, France, Germany, Italy, and the United States, who led the field in several technical areas [61].

Fig. 1.6. Paul Cornu's aircraft. Credits – Pilotfriend [135, 136].

In 1907, Paul Cornu constructed a vertical flight machine that was reported to have carried a human off the ground for the first time (see Figure 1.6). The airframe was very simple, with a rotor at each end. Power was supplied to the rotors by a 22 hp gasoline motor and belt transmission. Each rotor had two relatively large but low aspect ratio blades set at the periphery of a large spoked wheel. The rotors rotated in opposite directions to cancel torque reaction. A primitive means of control was achieved by placing auxiliary wings in the slipstream below the rotors. The flight lasted only twenty seconds and acquired an altitude of thirty centimeters but was still a landmark development in helicopter evolution. The helicopter had no effective means of control and was abandoned after a few flights [61, 64, 116].

By 1909, Igor Ivanovitch Sikorsky had built a nonpiloted coaxial helicopter prototype. This aircraft did not fly because of vibration problems and the lack of a powerful enough engine. In 1912, Boris Yur'ev tried to build a helicopter in Russia (Figure 1.7). This machine had a very modern looking single rotor and tail rotor configuration. The large diameter, high aspect ratio blades suggested some knowledge that this was the configuration for high aerodynamic efficiency. Yet, besides being one of the first to use a tail rotor design, Yur'ev was one of the firsts to propose the concepts of cyclic pitch for rotor control.

Fig. 1.7. Boris Yur'ev's aircraft [64, 148].

During World War I, military interest contributed to the advancement of the helicopter. Von Karman and Petrosczy, both from Germany, and the Hungarian Asboth intended to produce, without success, a lifting device to replace kite balloons for observation consisted of two superimposed lifting propellers (Figure 1.8).

Fig. 1.8. Petroczy – Von Karman's helicopter. Credits – Hiller Aviation Museum [66, 116].

It was not until late in World War I that major helicopter advances were made. The quality and quantity of production materials increased, and great improvements were made in the field of engine technology. With better technology and more need, the next step in helicopter advancement would soon come.

In 1922, George de Bothezat built a helicopter with four rotors, under the sponsorship of the U.S. Army (see Figure 1.9). The helicopter had four six-bladed rotors mounted at the ends of beams 18 m in length, forming a cross and intersecting in all directions. The rotor axes were not parallel but slightly inclined inwards so that if extended they would have met at a point directly above the centre of gravity. Besides the rotors with variable-pitch blades, the helicopter had two horizontal propellers called *steering airscrews* as well as two small airscrews placed above the gearbox and acting as regulators for the 220 hp engine. Ready for flight, the helicopter weighed 1700 kg.

In 1923, Juan de la Cierva developed the autogyro,[3] which resembled the helicopter, but used an unpowered rotor. This aircraft looked a lot like a hybrid between a fixed-wing airplane and a helicopter, with a set of conventional wings and a tail but with a rotor mounted on a vertical shaft above the fuselage (Figure 1.10). The blades were attached to the shaft for cyclic pitch control to balance the amount of lift and torque caused by the rotating blades and produce a stable ride. The articulated rotor blade is used today on all

---

[3] An autogyro is an aircraft with an unpowered rotary wing, or rotor, that resembles a helicopter. It is powered by either an engine-powered propeller or a tow cable. The movement of air past the rotor causes the lift [185].

Fig. 1.9. Bothezat's helicopter. Credits – National Museum of the United States Air Force [121].

helicopters. Two Cierva C.40 autogyros were used for Air Observation Post during World War I. Autogyros could neither hover nor descend vertically like the modern helicopter.

Fig. 1.10. Cierva's autogyro. Credits – Hiller Aviation Museum [66, 116].

In 1936, Heinrich Focke and Gert Achgelis built a side-by-side, two-rotor machine, called the *Fa-61* (Figure 1.11). This aerial vehicle was constructed from the fuselage of a small biplane trainer with rotor components provided by the *Weir-Cierva* company. Longitudinal control was achieved by tilting the rotors forward and aft by means of a swashplate mechanism, while yaw control was gained by tilting the rotors differentially. The rotors had no variable collective pitch, instead using a slow and clumsy system of changing rotor speed to change the rotor thrust. A vertical rudder and horizontal tail provided for additional directional stability. The cut-down propeller on the front of the machine served only to cool the radial engine. The *Fa-61* vehicle was the first helicopter to show fully controlled flight and also to demonstrate successful autorotations [61].

Fig. 1.11. The Fa-61 helicopter. Credits – Pilotfriend [64, 135, 136].

The success in the field of rotary-wing aviation was due almost entirely to Igor Sikorsky. In 1939, Sikorsky built the first *classical* helicopter, the VS-300 (Figure 1.12). This aircraft had one main rotor and three auxiliary tail rotors, with longitudinal and lateral control being obtained by means of pitch variations on the two vertically thrusting horizontal tail rotors. Powered only with a 75 hp engine, the machine could hover, fly sideways and backwards, and perform many other manoeuvres.

The main rotor of the *VS-300* was used in the *VS-300A* helicopter with a more powerful engine, but only the vertical (sideward thrusting) tail rotor was retained out of the original three auxiliary rotors. In this configuration, longitudinal and lateral control was achieved by tilting the main rotor by means of cyclic-pitch inputs; the single tail rotor was used for antitorque and directional control purposes. This configuration was to become the standard for most modern helicopters.

Fig. 1.12. Sikorsky's helicopters. Credits – Sikorsky Aircraft Corp. [64, 152].

During the 1950s many new advancements in helicopters were made. Sikorsky crafted the world's first certified commercial transport helicopter, the S-55 Chickasaw (H-19) (Figure 1.13).

Fig. 1.13. The S-55 helicopter. Credits – Enell Postcards [40].

Another well-known pioneer who contributed to the development of the modern helicopter was Stanley Hiller. In 1944, Hiller built the coaxial helicopter. His main breakthrough was the *rotormatic* main rotor design, where the cyclic pitch controls were connected to a set of small auxiliary blades set at ninety degrees to the main rotor blades [61].

The creation of the turbine engine advanced the helicopter's capabilities even further. With assembly lines brought about by the Industrial Revolution, these engines could be produced with high efficiency and increased precision.

At the beginning of the new millennium, the advance in many technologies, i.e. propulsion, materials, electronics, computers, sensors, navigation instruments, etc., contributed to the development of helicopters and other configurations of aerial vehicles capable of hovering.

The main roles of helicopters range from the civilian to the military. Civilian roles are surveillance, sea and mountain rescue, air ambulance, fire fighting, crop dusting, etc. Military roles include troop transport, mine-sweeping, battlefield surveillance, assault and anti-tank missions, etc.

Since the 1980s, sustained scientific research has been particularly interested in understanding the most difficult technical problems associated with helicopter flight, particularly in regard to aerodynamic limitations imposed by the main rotor. The improved design of the helicopter and the increasing viability of other vertical lift aircraft such as the tilt-rotor continue to advance as a result of the revolution in new technology.

The helicopter today is a safe, versatile, and reliable aircraft, that plays a unique role in modern aviation. The new generations of aircraft capable of hovering are designed to be smaller, lighter and with some autonomous functions.

1.3 Configuration of the Rotorcraft

Rotorcraft can be classified as follows:

1. Conventional main rotor/tail rotor configuration.

2. Single rotor configuration.

3. Twin rotor in coaxial configuration.

4. Twin rotor side by side.

5. Multi-rotor (example: four-rotor rotorcraft).

1.3.1 Conventional Main Rotor/Tail Rotor Configuration

The most common configuration is the combination of one main rotor and one tail rotor. The tail rotor compensates for the torque that is produced by the main rotor. The tail rotor also controls the helicopter along the vertical axis during hover flight (see Figure 1.14).

Conventional configuration has good controllability and manoeuvrability. However, the mechanical structure is complex and it requires a large rotor and a long tail boom.

Fig. 1.14. Conventional main rotor/tail rotor configuration.

1.3.2 Single Rotor Configuration

This type of aerodynamical configuration has a single rotor and ailerons to compensate the rotor torque (yaw control input). Since the rotor has no swash-plate, it has extra ailerons to produce pitch and roll torques (see Figure 1.15).

This type of flying machine is in general difficult to control even for experienced pilots.

Single rotor configuration is mechanically simpler than standard helicopters but it does not have as much control authority. In both cases a significant amount of energy is used in the anti-torque, i.e. to stop the fuselage turning around the vertical axis. Due to its mechanical simplicity, this configuration seems more suitable for micro-aircraft than the other configurations.

Fig. 1.15. Single rotor configuration – Eurosatory 2004. SAGEM, France.

1.3.3 Twin Rotor in Coaxial Configuration

In this configuration, one rotor is located on top of the other (see Figure 1.16). The two rotors turn in opposite directions. Depending on the angular velocity difference between the two rotors, the helicopter will turn left or right. Helicopters with this configuration cannot reach a high cruising speed because the drag is relatively large. Only after the development of the rigid rotor was it possible to build the two rotors closer together and reduce the drag considerably [116].

The *coaxial configuration* has advantages of compactness. However a significant amount of energy is lost because the rotors interfere with each other. The hover efficiency of the *twin rotor side by side* is higher than that of the coaxial because there is no interference between the rotors.

1.3.4 Twin Rotor Side by Side

The tandem rotor (or twin rotor side by side) configuration is used mainly in large helicopters (Figure 1.17). Because of the opposite rotation of the rotors,

Fig. 1.16. Twin rotor in coaxial configuration (AirScoot helicopter). Credits – AirScooter Corporation [2].

the torque of each single rotor is neutralized. The construction of the control system is much more complicated compared with a helicopter with a tail rotor. The arrangement of two rotors side by side was never very popular [116].

Fig. 1.17. Twin rotor side by side (Sea Knight CH-46 aircraft). Credits – National Museum of Naval Aviation [119].

1.3.5 Multi-rotors

The four-rotor rotorcraft or *quadrotor* is the more popular multi-rotor rotorcraft (Figure 1.18). This type of rotorcraft attempts to achieve stable hovering and precise flight by balancing the forces produced by the four rotors.

One of the advantages of using a multi-rotor helicopter is the increased payload capacity. It has more lift therefore heavier weights can be carried. Quadrotors are highly manoeuvrable, which enables vertical take-off and landing, as well as flying into hard to reach areas. Disadvantages are the increased helicopter weight and increased energy consumption due to the extra motors. Since it is controlled with rotor-speed changes, it is more suitable to electric motors, and large helicopter engines which have slow response may not be satisfactory without a proper gear-box system [5].

The *quadrotor* is superior from the control authority point of view. Controlled hover and low-speed flight has been successfully demonstrated. However further improvements are required before demonstrating sustained controlled forward-flight. When internal-combustion engines are used, multiple rotor configurations have disadvantages over single rotor configurations because of the complexity of the transmission gear.

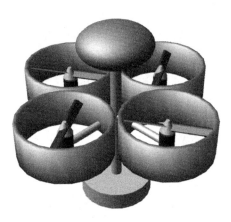

Fig. 1.18. Four-rotor rotorcraft (multi-rotor configuration).

1.4 New Configurations of UAVs

The aerodynamical configurations for mini-UAV are rapidly changing. In Figures 1.19 to 1.36 we present a selection of recent aerial vehicles capable of hovering [180].

Fig. 1.19. Seagull-Elbit Systems, Israel. Credits – Defense Update [31].

Fig. 1.20. Dragoneye-AeroVironment, Inc., U.S.A. Credits – Defense Update [31].

Fig. 1.21. Skylite-RAFAEL, Israel. Credits – Defense Update [31].

Fig. 1.22. Skylark-Elbit Systems, Israel. Credits – Defense Update [31].

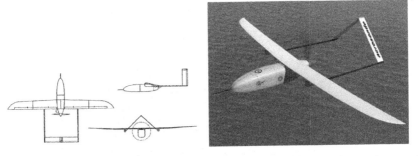

Fig. 1.23. Aerosonde aircraft - Aerosonde Robotic Aircraft. Credits – Aerosonde [1].

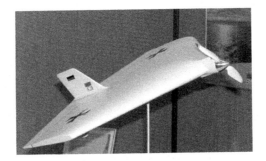

Fig. 1.24. Mikado aircraft - EMT, Germany. Credits – Defense Update [31].

Fig. 1.25. X-45 UCAV aircraft – Boeing Corp.

Fig. 1.26. Cypher aircraft. Credits – Sikorsky Aircraft Corp. [152].

Fig. 1.27. Cypher II aircraft. Credits – Sikorsky Aircraft Corp. [152].

Fig. 1.28. Golden Eye aircraft – Aurora Flight Sciences Corp.

Fig. 1.29. iSTAR MAV aircraft – Allied Aerospace. Credits – Defense Update [31].

Fig. 1.30. Kestrel aircraft – Honeywell.

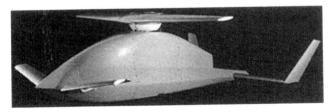

Fig. 1.31. X-50 aircraft – Boeing Corp.

Fig. 1.32. Guardian CL-327 aircraft – Bombadier Services Corp.

Fig. 1.33. TAGM65/M80 - Tactical Aerospace Group (TAG). Credits – Defense Update [31].

Fig. 1.34. T-Wing aircraft. Credits – University of Sydney, Australia [160].

Fig. 1.35. Four-rotor rotorcraft. Credits – Draganfly Innovations Inc. [33].

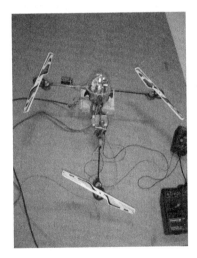

Fig. 1.36. Three-rotor rotorcraft. Credits – Heudiasyc-UTC, France

2

The PVTOL Aircraft

2.1 Introduction

We introduce in this chapter the well-known Planar Vertical Take-Off and Landing (PVTOL) aircraft problem. The PVTOL represents a challenging nonlinear systems control problem that is a particular case of what is today known as "motion control". The PVTOL is clearly motivated by the need to stabilize aircraft that are able to take-off vertically such as helicopters and some special airplanes.

The PVTOL is a mathematical model of a flying object that evolves in a vertical plane. It has three degrees of freedom (x, y, ϕ) corresponding to its position and orientation in the plane. The PVTOL is composed of two independent thrusters that produce a force and a moment on the flying machine, see Figure 2.1. The PVTOL is an underactuated system since it has three degrees of freedom and only two inputs [144]. The PVTOL is a very interesting nonlinear control problem.

Numerous design methods for the flight control of the PVTOL aircraft model exist in the literature [24, 26, 45, 53, 92, 96]. Indeed, this particular system is a simplified aircraft model with a minimal number of states and inputs but retains the main features that must be considered when designing control laws for a real aircraft. Since, the system possesses special properties such as, for instance, unstable zero dynamics and signed (thrust) input [63], several methodologies for controlling such a system have been proposed.

An algorithm to control the PVTOL based on an approximate I-O linearization procedure was proposed in [63]. Their algorithm achieves bounded tracking and asymptotic stability. A nonlinear small gain theorem was proposed in [173] which can be used to stabilize a PVTOL. The author has proved the stability of a controller based on nested saturations [172]. The algorithm is very simple, has bounded control inputs and performs well even if the initial horizontal displacement error is large.

An extension of the algorithm proposed by [63] was presented in [100]. They were able to find a flat output of the system that was used for tracking control of the PVTOL in presence of unmodelled dynamics.

The forwarding technique developed in [102] was used in [44] to propose a control algorithm for the PVTOL. This approach leads to a Lyapunov function which ensures asymptotic stability. Other techniques based on linearization were also proposed in [43].

Recently [99] proposed a control algorithm for the PVTOL for landing on a ship whose deck oscillates. They designed an internal model-based error feedback dynamic regulator that is robust with respect to uncertainties.

In this chapter we present a global stabilizing strategy for the control of the PVTOL aircraft. The stability proof is simple. The control algorithm has been tested in numerical simulations and in real-time applications.

This chapter is organized as follows. Section 2.2 presents the PVTOL dynamical model. Section 2.3.1 gives the control of the vertical displacement while the roll and horizontal displacement control are presented in Section 2.3.2. Real-time experimental results are given in Section 2.4. The conclusions are finally given in Section 2.5.

2.2 System Description

The dynamical model of the PVTOL aircraft can be obtained using the Lagrangian approach or Newton's laws. Therefore, the PVTOL system equations are given by (see Figure 2.1)

$$
\begin{aligned}
\ddot{x} &= -\sin(\phi)u_1 + \varepsilon\cos(\phi)u_2 \\
\ddot{y} &= \cos(\phi)u_1 + \varepsilon\sin(\phi)u_2 - 1 \\
\ddot{\phi} &= u_2
\end{aligned}
\tag{2.1}
$$

where x is the horizontal displacement, y is the vertical displacement and ϕ is the angle the PVTOL makes with the horizontal line. u_1 is the total thrust and u_2 is the couple as shown in Figure 2.1. The parameter ε is a small coefficient which characterizes the coupling between the rolling moment and the lateral acceleration of the aircraft. The constant -1 is the normalized gravitational acceleration.

In general, ε is negligible and not always well known [63]. Therefore, it is possible to suppose that $\varepsilon = 0$, i.e.

$$
\begin{aligned}
\ddot{x} &= -\sin(\phi)u_1 \\
\end{aligned}
\tag{2.2}
$$

$$
\ddot{y} = \cos(\phi)u_1 - 1 \tag{2.3}
$$

$$
\ddot{\phi} = u_2 \tag{2.4}
$$

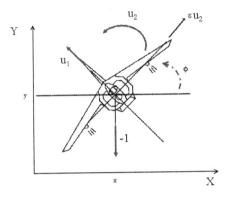

Fig. 2.1. The PVTOL aircraft (front view).

Furthermore, several authors have shown that by an appropriate change
of coordinates, we can obtain a representation of the system without the term
due to ε [125]. For instance, R. Olfati-Saber [125] applied the following change
of coordinates

$$\bar{x} = x - \varepsilon \sin(\phi) \tag{2.5}$$
$$\bar{y} = y + \varepsilon(\cos(\phi) - 1) \tag{2.6}$$

The system dynamics considering these new coordinates become

$$\begin{aligned}
\ddot{\bar{x}} &= -\sin(\phi)\bar{u}_1 \\
\ddot{\bar{y}} &= \cos(\phi)\bar{u}_1 - 1 \\
\ddot{\phi} &= u_2
\end{aligned} \tag{2.7}$$

where $\bar{u}_1 = u_1 - \varepsilon \dot{\phi}^2$. Note that this structure (2.7) has the same form as (2.1)
with $\varepsilon = 0$.

2.3 Control Strategy

In this section we present a simple control algorithm for the PVTOL whose
convergence analysis is also relatively simple as compared with other controllers proposed in the literature. We present a new approach based on nested
saturations to control the PVTOL which can lead to further developments in
nonlinear systems. The simplicity of both the algorithm and the analysis allows for a better understanding of the problem. The proposed algorithm can
be considered as an extension of the control of multiple integrators of [172] to
the case when there are nonlinear functions between the integrators.

This section is divided in two parts. In the first part, we are interested in stabilizing the altitude y, while in the second part we propose u_2 in order to control the roll angle and the horizontal displacement in the x-axis.

2.3.1 Control of the Vertical Displacement

The vertical displacement y will be controlled by forcing the altitude to behave as a linear system. This is done by using the following control strategy

$$u_1 = \frac{r_1 + 1}{\cos \sigma_p(\phi)} \tag{2.8}$$

where $0 < p < \frac{\pi}{2}$ and σ_η, for some $\eta > 0$, is a saturation function

$$\sigma_\eta(s) = \begin{cases} \eta & \text{for } s > \eta \\ s & \text{for } -\eta \leq s \leq \eta \\ -\eta & \text{for } s < -\eta \end{cases} \tag{2.9}$$

and

$$r_1 = -a_1 \dot{y} - a_2(y - y_d) \tag{2.10}$$

where y_d is the desired altitude and a_1 and a_2 are positive constants such that the polynomial $s^2 + a_1 s + a_2$ is stable. Let us assume that after a finite time T_2, $\phi(t)$ belongs to the interval

$$I_{\frac{\pi}{2}} = (-\frac{\pi}{2} + \epsilon, \frac{\pi}{2} - \epsilon) \tag{2.11}$$

for some $\epsilon > 0$ so that $\cos \phi(t) \neq 0$. Introducing (2.8) and (2.10) into (2.2)–(2.4) we obtain for $t > T_2$

$$\begin{aligned} \ddot{x} &= -\tan \phi(r_1 + 1) \\ \ddot{y} &= -a_1 \dot{y} - a_2(y - y_d) \\ \ddot{\phi} &= u_2 \end{aligned} \tag{2.12}$$

Note that in view of the above, $y \to y_d$ and $r_1 \to 0$ as $t \to \infty$.

2.3.2 Control of the Roll Angle and the Horizontal Displacement

We will now propose u_2 to control $\dot{\phi}, \phi, \dot{x}$ and x. The control algorithm will be obtained step by step. The final expression for u_2 will be given at the end of this section (see (2.63)). Roughly speaking, for ϕ close to zero, the (x, ϕ) subsystem is represented by four integrators in cascade. A. Teel [172] proposed a control strategy based on nested saturations that can be used to control a set of integrators connected in cascade. We will show that the strategy proposed in [172] for linear systems can be modified to control x and ϕ in the nonlinear system (2.12).

We will also show that $\phi(t) \in I_{\frac{\pi}{2}}$ in (2.11) after $t = T_2$ independently of the input u_1 in (2.8).

Boundedness of $\dot{\phi}$

In order to establish a bound for $\dot{\phi}$ define u_2 as

$$u_2 = -\sigma_a(\dot{\phi} + \sigma_b(z_1)) \tag{2.13}$$

where $a > 0$ is the desired upper bound for $|u_2|$ and z_1 will be defined later. Let

$$V_1 = \tfrac{1}{2}\dot{\phi}^2 \tag{2.14}$$

Then it follows that

$$\dot{V}_1 = -\dot{\phi}\sigma_a(\dot{\phi} + \sigma_b(z_1)) \tag{2.15}$$

Note that if $|\dot{\phi}| > b + \delta$ for some $b > 0$ and some $\delta > 0$ arbitrarily small, then $\dot{V}_1 < 0$. Therefore, after some finite time T_1, we will have

$$|\dot{\phi}(t)| \leq b + \delta \tag{2.16}$$

Let us assume that b satisfies

$$a \geq 2b + \delta \tag{2.17}$$

Then, from (2.12) and (2.13) we obtain for $t \geq T_1$

$$\ddot{\phi} = -\dot{\phi} - \sigma_b(z_1) \tag{2.18}$$

Boundedness of ϕ

To establish a bound for ϕ, define z_1 as

$$z_1 = z_2 + \sigma_c(z_3) \tag{2.19}$$

for some z_3 to be defined later and

$$z_2 = \phi + \dot{\phi} \tag{2.20}$$

From (2.18)–(2.20) we have

$$\dot{z}_2 = -\sigma_b(z_2 + \sigma_c(z_3)) \tag{2.21}$$

Let

$$V_2 = \tfrac{1}{2}z_2^2 \tag{2.22}$$

then

$$\dot{V}_2 = -z_2\sigma_b(z_2 + \sigma_c(z_3)) \tag{2.23}$$

Note that if $|z_2| > c + \delta$ for some δ arbitrarily small and some $c > 0$, then $\dot{V}_2 < 0$. Therefore, it follows that after some finite time $T_2 \geq T_1$, we have

$$|z_2(t)| \leq c + \delta \tag{2.24}$$

From (2.20) we obtain for $t \geq T_2$

$$\phi(t) = \phi(T_2)e^{-(t-T_2)} + \int_{T_2}^{t} e^{-(t-\tau)} z_2(\tau) d\tau \tag{2.25}$$

Therefore, it follows that there exists a finite time T_3 such that for $t \geq T_3 > T_2$ we have

$$|\phi(t)| \leq \bar{\phi} \triangleq c + 2\delta \tag{2.26}$$

If

$$c + 2\delta \leq \frac{\pi}{2} - \epsilon \tag{2.27}$$

then $\phi(t) \in I_{\frac{\pi}{2}}$, see (2.11), for $t \geq T_2$.

Assume that b and c also satisfy

$$b \geq 2c + \delta \tag{2.28}$$

Then, in view of (2.24), (2.21) reduces to

$$\dot{z}_2 = -z_2 - \sigma_c(z_3) \tag{2.29}$$

for $t \geq T_3$.

Note that the following inequality holds for $|\phi| < 1$

$$|\tan \phi - \phi| \leq \phi^2 \tag{2.30}$$

We will use the above inequality in the following development.

Boundedness of $\dot{x}$

To establish a bound for $\dot{x}$, let us define z_3 as

$$z_3 = z_4 + \sigma_d(z_5) \tag{2.31}$$

where z_4 is defined as

$$z_4 = z_2 + \phi - \dot{x} \tag{2.32}$$

and z_5 will be defined later. From (2.12), (2.20) and (2.29) and the above it follows that

$$\dot{z}_4 = (1 + r_1) \tan \phi - \phi - \sigma_c(z_4 + \sigma_d(z_5)) \tag{2.33}$$

Define

$$V_3 = \tfrac{1}{2} z_4^2 \tag{2.34}$$

then

$$\dot{V}_3 = z_4 \left[(1 + r_1) \tan \phi - \phi - \sigma_c(z_4 + \sigma_d(z_5)) \right] \tag{2.35}$$

Since $r_1 \tan \phi \to 0$ (see (2.10) and (2.12)), there exists a finite time $T_5 > T_4$, large enough such that if

$$|z_4| > d + \bar{\phi}^2 + \delta \tag{2.36}$$

and

$$c \geq \bar{\phi}^2 + \delta \tag{2.37}$$

for some δ arbitrarily small and $d > 0$, then $\dot{V}_3 < 0$. Therefore, after some finite time $T_6 > T_5$, we have

$$|z_4(t)| \leq d + \delta + \bar{\phi}^2 \tag{2.38}$$

Let us assume that d and c satisfy

$$c \geq 2d + \delta + \bar{\phi}^2 \tag{2.39}$$

Thus, after a finite time T_6, (2.33) reduces to

$$\dot{z}_4 = (1 + r_1) \tan \phi - \phi - z_4 - \sigma_d(z_5) \tag{2.40}$$

Note that in view of (2.20), (2.32) and (2.38) it follows that $\dot{x}$ is bounded.

Boundedness of x

To establish a bound for x, let us define z_5 as

$$z_5 = z_4 + \phi - 2\dot{x} - x \tag{2.41}$$

From (2.12), (2.20), (2.32) and (2.40) we get

$$\dot{z}_5 = (1 + r_1) \tan \phi - \phi - z_4 - \sigma_d(z_5) + \dot{\phi} + 2 \tan \phi(r_1 + 1) - \dot{x}$$
$$= -\sigma_d(z_5) + 3r_1 \tan \phi + 3(\tan \phi - \phi) \tag{2.42}$$

Define

$$V_4 = \tfrac{1}{2} z_5^2 \tag{2.43}$$

then

$$\dot{V}_4 = z_5 \left[-\sigma_d(z_5) + 3r_1 \tan \phi + 3(\tan \phi - \phi) \right] \tag{2.44}$$

Since $r_1 \tan \phi \to 0$, there exists a finite time $T_7 > T_6$, large enough such that if $|z_5| > 3\bar{\phi}^2 + \delta$ for some δ arbitrarily small and

$$d \geq 3\bar{\phi}^2 + \delta \tag{2.45}$$

then $\dot{V}_4 < 0$. Therefore, after some finite time $T_8 > T_7$, we have

$$|z_5(t)| \leq 3\bar{\phi}^2 + \delta \tag{2.46}$$

After time T_8 (2.42) reduces to

$$\dot{z}_5 = -z_5 + 3r_1 \tan \phi + 3(\tan \phi - \phi) \tag{2.47}$$

Boundedness of x follows from (2.38), (2.41) and (2.46).

Let us rewrite all the constraints on the parameters a, b, c, d and $\bar{\phi}$

$$a \geq 2b + \delta \tag{2.48}$$
$$\bar{\phi} \triangleq c + 2\delta \leq 1 \tag{2.49}$$
$$b \geq 2c + \delta \tag{2.50}$$
$$c \geq (c + 2\delta)^2 + 2d + \delta \tag{2.51}$$
$$d \geq 3(c + 2\delta)^2 + \delta \tag{2.52}$$

From the above we obtain

$$a \geq 4c + 3\delta \tag{2.53}$$
$$b \geq 2c + \delta \tag{2.54}$$
$$c + 2\delta \leq 1 \tag{2.55}$$
$$\begin{aligned} c &\geq (c + 2\delta)^2 + 2d + \delta \\ &\geq (c + 2\delta)^2 + 2(3(c + 2\delta)^2 + \delta) + \delta \\ &\geq 7(c + 2\delta)^2 + 3\delta \end{aligned} \tag{2.56}$$
$$d \geq 3(c + 2\delta)^2 + \delta \tag{2.57}$$

Convergence of ϕ, $\dot{\phi}$, x and $\dot{x}$ to zero

Therefore, c and δ should be chosen small enough to satisfy (2.53) and (2.57). The parameters a, b and d can then be computed as a function of c as above.

From (2.47) it follows that for a time large enough,

$$\mid z_5(t) \mid \leq 3\phi^2 + \delta \tag{2.58}$$

for some δ arbitrarily small. From (2.40) and (2.58) we have that for a time large enough,

$$\mid z_4(t) \mid \leq 4\phi^2 + 2\delta \tag{2.59}$$

for some δ arbitrarily small. From (2.31) and the above we have

$$\mid z_3(t) \mid \leq 7\phi^2 + 3\delta \tag{2.60}$$

Similarly, from (2.29)

$$\mid z_2(t) \mid \leq 7\phi^2 + 4\delta \tag{2.61}$$

and finally for a time large enough and an arbitrarily small δ, from (2.25) and the above we get

$$\mid \phi \mid \leq 7\phi^2 + 5\delta \tag{2.62}$$

Since δ is arbitrarily small, the above inequality implies that

$$\phi = 0$$
$$|\phi| \geq \frac{1}{7}$$

If c is chosen small enough such that $\bar{\phi} < \frac{1}{7}$ (see (2.26)), then the only possible solution is $\phi = 0$. Therefore $\phi \to 0$ as $t \to \infty$. From (2.58)–(2.61) and (2.19) we have that $z_i(t) \to 0$ for $i = 1, 2, ..., 5$. From (2.20) we get $\dot{\phi} \to 0$. From (2.32) and (2.41) it follows respectively that $\dot{x} \to 0$ and $x \to 0$. The control input u_2 is given by (2.13), (2.19), (2.20), (2.31), (2.32) and (2.41), i.e.

$$u_2 = -\sigma_a(\dot{\phi} + \sigma_b(\phi + \dot{\phi} + \sigma_c(2\phi + \dot{\phi} - \dot{x} + \sigma_d(3\phi + \dot{\phi} - 3\dot{x} - x)))) \quad (2.63)$$

The amplitudes of the saturation functions should satisfy the constraints in (2.53)–(2.57).

2.4 Real-time Experimental Results

In this section we present a real-time application of the control algorithm for the PVTOL presented in the previous sections. We have tested our control algorithm in two experiments, the first one was made using a quad-rotor rotorcraft and the second one with a prototype of the PVTOL built at the Heudiasyc laboratory.

Fig. 2.2. Photograph of the quad-rotor rotorcraft (front view).

2.4.1 Experiment using a Quad-Rotor Rotorcraft

We use a four-rotor electric mini-helicopter as shown in Figure 2.2. Note that when the yaw and pitch angles are set to zero, the quad-rotor rotorcraft reduces to a PVTOL.

In our experiment the pitch and yaw angles are controlled manually by an experienced pilot. The remaining controls, i.e. the main thrust and the roll control, are controlled using the control strategy presented in the previous sections. In the four-rotor helicopter, the main thrust is the sum of the thrusts of each motor. Pitch movement is obtained by increasing (reducing) the speed of the rear motor while reducing (increasing) the speed of the front motor. The roll movement is obtained similarly using the lateral motors.

The radio is a Futaba Skysport 4. The radio and the PC (INTEL Pentium III) are connected using data acquisition cards (ADVANTECH PCL-818HG and PCL-726). The connection in the radio is directly made to the joystick potentiometers for the main thrust and pitch control.

We use the 3D tracker system (POLHEMUS) [46] for measuring the position and the orientation of the rotorcraft. The Polhemus is connected via RS232 to the PC.

In Chapter 3, we will give a detailed description of this platform.

We wish to use our control law with a quad-rotor rotorcraft; this helicopter evolves in 3D and its movements are defined by the variables $(x, y, z, \psi, \theta, \phi)$. We are going to assimilate the altitude of the quad-rotor rotorcraft to the altitude of the PVTOL. This means that we will see y of the PVTOL as z of the quad-rotor rotorcraft.

The control objective for this experience is to make the quad-rotor rotorcraft hover at an altitude of 20 cm i.e. we wish to reach the position $(x, z) = (0, 20)$ cm while $\phi = 0°$.

With respect to the time derivatives, we have simplified the mathematical computation by setting $\dot{q}_t \cong \frac{q_t - q_{t-T}}{T}$, where q is a given variable and T is the sampling period. In the experiment $T = 71$ ms.

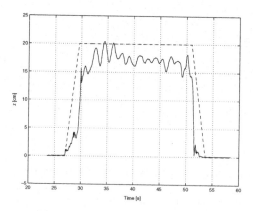

Fig. 2.3. Altitude of the quad-rotor rotorcraft.

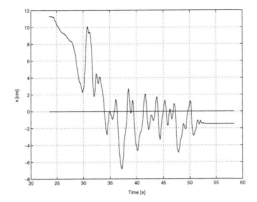

Fig. 2.4. x position of the quad-rotor rotorcraft.

Figures 2.3–2.4 show the performance of the controller when applied to the helicopter. Take-off and landing were performed autonomously. The choice of the values for a, b, c, d were carried out satisfying inequalities $(2.53)-(2.57)$. However these parameters have been tuned experimentally in the sequence as they appear in the control input u_2.

Figure 2.3 describes the gap between the real altitude of the rotorcraft and the desired altitude. One can notice that it follows satisfactorily the desired reference. Following the vertical axis, we notice that the position error in the variable z is of about 3 to 4 cm. In Figure 2.4, one can see the gap between the real horizontal position of the quad-rotor rotorcraft (according to x) and the desired position. We note that the rotorcraft position converges towards the desired position. The final position error is about 2 to 4 cm.

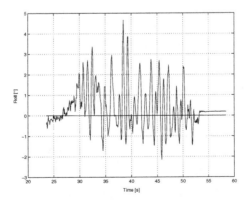

Fig. 2.5. Roll angle (ϕ) of the quad-rotor rotorcraft.

Figure 2.5 shows the evolution of the ϕ angle. It can be seen that the control law achieves convergence of the roll angle to zero, as the helicopter goes up to the altitude of 20 cm. It also shows the rotorcraft horizontal displacement in the x-axis.

2.4.2 Experimental Platform Using Vision

In this subsection we have tested the control algorithm using vision [131].

The PVTOL prototype built at Heudiasyc laboratory is shown in Figure 2.6. The rotors are driven separately by two motors. One motor rotates clockwise while the second one rotates counter-clockwise. The main thrust is the sum of the thrusts of each motor. The rolling moment is obtained by increasing (decreasing) the speed of one motor while decreasing (increasing) the speed of the second motor.

Fig. 2.6. Photograph of the PVTOL aircraft (front view).

The PVTOL moves on an inclined plane, which defines our 2D workspace. The PVTOL platform is an experimental setup designed to study the problems currently found in navigation at low altitude of a small flying object. At low altitude, GPS and even intertial navigation systems are not enough to stabilize mini-flying objects. Vision using cameras should provide additional information to make autonomous flights near the ground possible. For simplicity, at a first stage, we have placed the camera outside the aircraft. In the future, the camera will be located at the base of the mini-helicopter, pointing downwards. Note that even when the camera is located outside the flying object, we still have to deal with the problems of object localization computation using cameras and delays in the closed-loop system.

In the platform, a CCD camera Pulnix is located perpendicular to the plane at a fixed altitude and provides an image of the whole workspace. We have used an acquisition card PCI-1409 of National Instruments Company. The camera is linked to the PC dedicated to the vision part (which will be referred as Vision PC). From the image provided by the camera, the program calculates the position (x, y) and the orientation ϕ of the PVTOL with respect to a given origin. Then, the Vision PC sends this information to another PC dedicated to the control part (which we call Control PC), via a RS232 connection, transmitting at 115 200 bps (bauds per second).

The control inputs are therefore calculated according to the proposed strategy based on saturation functions and sent to the PVTOL via the radio. In order to simplify the implementation of the control law we have designed the platform in such a way that each of the two control inputs can independently work either in automatic or in manual mode. The minimum sampling period we are able to obtain in the experimental platform is 40 ms. This includes the computation of the control law, image processing, localization computation and A/D and D/A conversion in the radio–PC interface.

Experiment

The control objective is to make the PVTOL hover at an altitude of 60 pixels i.e. we wish to reach the position $(x, y) = (0, 60)$ in pixels while $\phi = 0°$.

The measurements of x, y are expressed in pixels in the image frame and ϕ is expressed in degrees, which means that the servoing is done on the basis of image features directly.

With respect to the time derivatives, we have simplified the mathematical computation by setting $\dot{q}_t = \frac{q_t - q_{t-T}}{T}$, where q represent either x, y or ϕ and T is the sampling period.

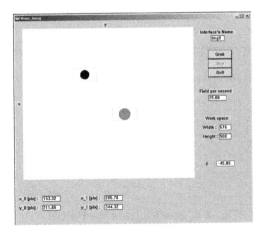

Fig. 2.7. Vision interface.

In Figure 2.7, the results of the image acquisition program are shown. We clearly see the detection of two points located on the PVTOL prototype. From the measurement of these two points, we compute the position x, y and the angle ϕ of the system.

Figures 2.8–2.10 show the performance of the controller when applied to the PVTOL aircraft. Hovering at 60 pixels as well as following a horizontal trajectory were performed satisfactorily. The choice of the values for a, b, c, d were carried out satisfying inequalities (2.53)–(2.57). However these parameters have been tuned experimentally in the sequence as they appear in the control input u_2.

In Figure 2.8, one can see the gap between the real horizontal position of the PVTOL in the x-axis and the desired position. In the x-axis, 1 cm corresponds to 5 pixels. This means that the error position is about 2 to 3 cm.

Figure 2.9 presents the gap between the real PVTOL altitude and the desired altitude. One can notice that the PVTOL follows the desired reference. In the vertical y-axis, 1 cm corresponds to 2.5 pixels. This means that the vertical position error is about 2 to 3 cm.

Figure 2.10 shows the evolution of the ϕ-angle In this figure, one can see that the control law achieves convergence of the ϕ-angle to zero. Therefore, the difference between the real trajectory and desired one is due to friction. The results are nevertheless very satisfactory.

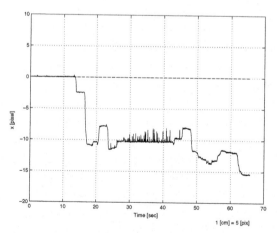

Fig. 2.8. Position $[x]$ of the PVTOL.

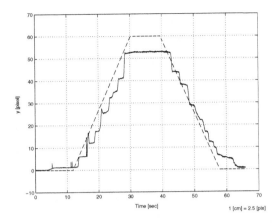

Fig. 2.9. Position $[y]$ of the PVTOL.

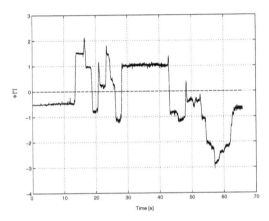

Fig. 2.10. Orientation $[\phi]$ of the PVTOL.

2.5 Conclusion

We have presented a simple control algorithm for stabilizing the PVTOL. The controller is an extension of the nested saturations technique introduced in [172]. The convergence Lyapunov analysis has shown that the proposed algorithm is asymptotically stable.

We have been able to test the algorithm in real-time applications. The simplicity of the algorithm was very useful in the implementation of the control algorithm. The results showed that the algorithm performs well. We were able to perform autonomously the tasks of take-off, hover and landing.

This platform exhibits the same difficulties found in autonomous flight close to the ground and can be used as a benchmark for developing controllers for unmanned flying vehicles.

3

The Quad-rotor Rotorcraft

3.1 Introduction

The helicopter is one of the most complex flying machines. Its complexity is due to its versatility and manoeuvrability to perform many types of tasks [12], [55]. The classical helicopter is conventionally equipped with a main rotor and a tail rotor. However other types of helicopters exist including the twin rotor or tandem helicopter and the co-axial rotor helicopter. In this chapter we are particularly interested in controlling a mini-rotorcraft having four rotors. This rotorcraft is also known as a quad-rotor rotorcraft.

The quad-rotor rotorcraft is not a new configuration, it already existed in the year 1922. In January 1921, the US Army Air Corps awarded a contract to Dr. George de Bothezat and Ivan Jerome to develop a vertical flight machine. The 1678 kg X-shaped structure supported a 8.1 m diameter six-blade rotor at each end of the 9 m arms (see Figure 3.1). At the ends of the lateral arms, two small propellers with variable pitch were used for thrusting and yaw control. A small lifting rotor was also mounted above the 180 hp *Le Rhone* radial engine at the junction of the frames, but was later removed as unnecessary [74].

Each rotor had individual collective pitch control to produce differential thrust through vehicle inclination for translation. The aircraft weighed 1700 kg at take-off and made its first flight in October 1922. The engine was soon upgraded to a 220 hp Bentley BR-2 rotary. About 100 flights were made by the end of 1923 at what would eventually be known as Wright Field near Dayton, Ohio, including one with three passengers hanging onto the airframe. Although the contract called for a 100 m hover, the highest it ever reached was about 5 m. After expending $200,000, de Bothezat demonstrated that his vehicle could be quite stable and that the practical helicopter was theoretically possible. It was, however, underpowered, unresponsive, mechanically complex and susceptible to reliability problems. Pilot workload was too high during hover to attempt lateral motion.

Fig. 3.1. The quad-rotor rotorcraft of Bothezat.

Few works are reported in the literature for the helicopter having four rotors. Indeed, it was only last year that the interest of researchers and specialists of aeronautics increased considerably.

Young et al. [188] sponsored by the Directorate Aerospace in NASA Ames Research Center present new configurations of mini-drones and their applications among which is the helicopter with four rotors called the Quad-Rotor-Tail-Sitter.

Pounds et al. [139] conceived and developed a control algorithm for a prototype of an aerial vehicle having four rotors. They considered using an MIU (Measurement Inertial Unit) to measure the speed and the angular acceleration. They use a linearization of the dynamic model to conceive the control algorithm. The results of the control law have been tested in simulation.

Altuğ et al. [5] proposed a control algorithm to stabilize the quad-rotor using vision as the principal sensor. They studied two methods, the first uses a control algorithm of linearization and the other uses the technique of backstepping. They have tested the control laws in simulation. They also present an experience using vision to measure the yaw angle and the altitude.

In this chapter we present the model of a quad-rotor rotorcraft whose dynamical model is obtained via a Lagrange approach [25, 94]. A control strategy is proposed having in mind that the quad-rotor rotorcraft can be seen as the interconnection of two PVTOL aircraft. Indeed, we first design a control to stabilize the yaw angular displacement. We then control the pitch movement using a controller based on the dynamic model of a PVTOL (see [63]). Finally the roll movement is controlled using again a strategy based on the PVTOL.

The control algorithm is based on the nested saturation control strategy proposed by [172] and discussed for general nonlinear systems, including the PVTOL, in [173]. We prove global stability of the proposed controller. Furthermore the controller has been implemented on a PC and real-time experiments have shown that the proposed control strategy performs well in practice. Robustness with respect to parameter uncertainty and unmodelled dynamics has been observed in a real-time application performed by the aerial control team at the University of Technology of Compiègne, France.

The chapter is organized as follows. The principal characteristics of the quad-rotor rotorcraft and the dynamical model are given in Section 3.2. Section 3.3 describes the control law design. In Section 3.4 we present the experimental results and finally, some conclusions are given in Section 3.5.

3.2 Dynamic Model

In this section we present the principal characteristics of the quad-rotor rotorcraft and the dynamic model of the quad-rotor rotorcraft using a Lagrange approach.

3.2.1 Characteristics of the Quad-rotor Rotorcraft

Quad-rotor rotorcraft, like the one shown in Figure 3.2, have some advantages over conventional helicopters. Given that the front and the rear motors rotate counter-clockwise while the other two rotate clockwise, gyroscopic effects and aerodynamic torques tend to cancel in trimmed flight.

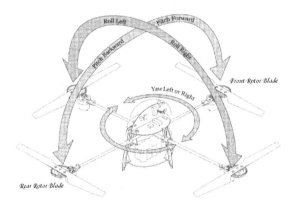

Fig. 3.2. The quad-rotor rotorcraft.

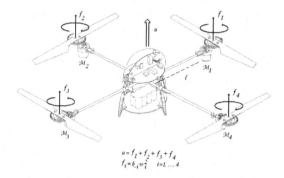

$$u = f_1 + f_2 + f_3 + f_4$$
$$f_i = k_i u_i^2 \quad i = 1, \dots, 4$$

Fig. 3.3. The throttle control input.

This four-rotor rotorcraft does not have a swashplate. In fact it does not need any blade pitch control. The collective input (or throttle input) is the sum of the thrusts of each motor (see Figure 3.3). Pitch movement is obtained by increasing (reducing) the speed of the rear motor while reducing (increasing) the speed of the front motor. The roll movement is obtained similarly using the lateral motors. The yaw movement is obtained by increasing (decreasing) the speed of the front and rear motors while decreasing (increasing) the speed of the lateral motors. This should be done while keeping the total thrust constant.

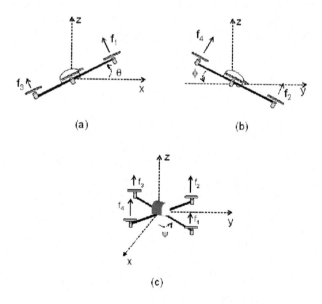

Fig. 3.4. (a) Pitch, (b) roll and (c) yaw angles.

In view of its configuration, the quad-rotor rotorcraft in Figure 3.2 has some similarities with the PVTOL (Planar Vertical Take Off and Landing) aircraft problem [43], [99]. Furthermore, the quad-rotor rotorcraft reduces to a PVTOL when the pitch and yaw angles are set to zero (see Section 2.4.1). In a way the quad-rotor rotorcraft can be seen as two PVTOLs connected such that their axes are orthogonal (see Figure 3.4 (a) and (b)).

3.2.2 System Description

In this subsection we will describe the dynamical model we have used for the quad-rotor mini-rotorcraft. This model is basically obtained by representing the mini-rotorcraft as a solid body evolving in 3D and subject to one force and three moments. The four electric motors' dynamics is relatively fast and therefore it will be neglected as well as the flexibility of the blades.

The generalized coordinates for the rotorcraft are

$$q = (x, y, z, \psi, \theta, \phi) \in R^6 \tag{3.1}$$

where (x, y, z) denote the position of the centre of mass of the four-rotor rotorcraft relative to the frame $\mathcal{I}$, and (ψ, θ, ϕ) are the three Euler angles (yaw, pitch and roll angles) and represent the orientation of the rotorcraft [41], [4].

Therefore, the model partitions naturally into translational and rotational coordinates

$$\xi = (x, y, z) \in \Re^3, \qquad \eta = (\psi, \theta, \phi) \in \mathcal{S}^3 \tag{3.2}$$

The translational kinetic energy of the rotorcraft is

$$T_{trans} \triangleq \frac{m}{2} \dot{\xi}^T \dot{\xi} \tag{3.3}$$

where m denotes the mass of the rotorcraft. The rotational kinetic energy is

$$T_{rot} \triangleq \frac{1}{2} \dot{\eta}^T \mathbb{J} \dot{\eta} \tag{3.4}$$

The matrix $\mathbb{J}$ acts as the inertia matrix for the full rotational kinetic energy of the rotorcraft expressed directly in terms of the generalized coordinates η. The only potential energy which needs to be considered is the standard gravitational potential given by

$$U = mgz \tag{3.5}$$

The Lagrangian is

$$
\begin{aligned}
L(q, \dot{q}) &= T_{trans} + T_{rot} - U \\
&= \frac{m}{2} \dot{\xi}^T \dot{\xi} + \frac{1}{2} \dot{\eta}^T \mathbb{J} \dot{\eta} - mgz
\end{aligned}
\tag{3.6}
$$

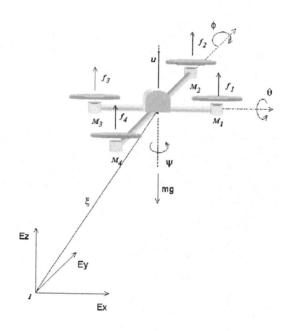

Fig. 3.5. The quad-rotor in an inertial frame.

The model for the full rotorcraft dynamics is obtained from the Euler–Lagrange equations with external generalized force

$$\frac{d}{dt}\frac{\partial \mathcal{L}}{\partial \dot{q}} - \frac{\partial \mathcal{L}}{\partial q} = F \qquad (3.7)$$

where $F = (F_\xi, \tau)$. τ is the generalized moments and F_ξ is the translational force applied to the rotorcraft due to the control inputs. We ignore the small body forces because they are generally of a much smaller magnitude than the principal control inputs u and τ, then we write

$$\widehat{F} = \begin{pmatrix} 0 \\ 0 \\ u \end{pmatrix} \qquad (3.8)$$

where (see Figures 3.5 and 3.3)

$$u = f_1 + f_2 + f_3 + f_4 \qquad (3.9)$$

and

$$f_i = k_i w_i^2, \qquad i = 1, ..., 4 \tag{3.10}$$

where $k_i > 0$ is a constant and w_i is the angular speed of motor "i" (M_i, $i = 1, ..., 4$), then

$$F_\xi = R\widehat{F} \tag{3.11}$$

where R is the transformation matrix representing the orientation of the rotorcraft, we use c_θ for $\cos\theta$ and s_θ for $\sin\theta$.

$$R = \begin{pmatrix} c_\theta c_\psi & s_\psi s_\theta & -s_\theta \\ c_\psi s_\theta s_\phi - s_\psi c_\phi & s_\psi s_\theta s_\phi + c_\psi c_\phi & c_\theta s_\phi \\ c_\psi s_\theta c_\phi + s_\psi s_\phi & s_\psi s_\theta c_\phi - c_\psi s_\phi & c_\theta c_\phi \end{pmatrix} \tag{3.12}$$

The generalized moments on the η variables are

$$\tau \triangleq \begin{pmatrix} \tau_\psi \\ \tau_\theta \\ \tau_\phi \end{pmatrix} \tag{3.13}$$

where

$$\tau_\psi = \sum_{i=1}^{4} \tau_{M_i}$$
$$\tau_\theta = (f_2 - f_4)\ell$$
$$\tau_\phi = (f_3 - f_1)\ell$$

where ℓ is the distance from the motors to the centre of gravity and τ_{M_i} is the couple produced by motor M_i.

Since the Lagrangian contains no cross-terms in the kinetic energy combining $\dot{\xi}$ and $\dot{\eta}$ (see (3.6)), the Euler–Lagrange equation can be partitioned into the dynamics for the ξ coordinates and the η dynamics. One obtains

$$m\ddot{\xi} + \begin{pmatrix} 0 \\ 0 \\ mg \end{pmatrix} = F_\xi \tag{3.14}$$

$$\mathbb{J}\ddot{\eta} + \dot{\mathbb{J}}\dot{\eta} - \frac{1}{2}\frac{\partial}{\partial\eta}\left(\dot{\eta}^T \mathbb{J}\dot{\eta}\right) = \tau \tag{3.15}$$

Defining the Coriolis/Centripetal vector

$$\bar{V}(\eta, \dot{\eta}) = \dot{\mathbb{J}}\dot{\eta} - \frac{1}{2}\frac{\partial}{\partial\eta}\left(\dot{\eta}^T \mathbb{J}\dot{\eta}\right) \tag{3.16}$$

we may write

$$\mathbb{J}\ddot{\eta} + \bar{V}(\eta, \dot{\eta}) = \tau \tag{3.17}$$

but we can rewrite $\bar{V}(\eta, \dot{\eta})$ as

$$\bar{V}(\eta, \dot{\eta}) = (\dot{\mathbb{J}} - \frac{1}{2}\frac{\partial}{\partial \eta}(\dot{\eta}^T \mathbb{J}))\dot{\eta}$$
$$= C(\eta, \dot{\eta})\dot{\eta} \qquad (3.18)$$

where $C(\eta, \dot{\eta})$ is referred to as the Coriolis terms and contains the gyroscopic and centrifugal terms associated with the η dependence of $\mathbb{J}$.

Finally we obtain

$$m\ddot{\xi} = u \begin{pmatrix} -\sin\theta \\ \cos\theta \sin\phi \\ \cos\theta \cos\phi \end{pmatrix} + \begin{pmatrix} 0 \\ 0 \\ -mg \end{pmatrix} \qquad (3.19)$$

$$\mathbb{J}\ddot{\eta} = -C(\eta, \dot{\eta})\dot{\eta} + \tau \qquad (3.20)$$

In order to simplify let us propose a change of the input variables.

$$\tau = C(\eta, \dot{\eta})\dot{\eta} + \mathbb{J}\tilde{\tau} \qquad (3.21)$$

where

$$\tilde{\tau} = \begin{pmatrix} \tilde{\tau}_\psi \\ \tilde{\tau}_\theta \\ \tilde{\tau}_\phi \end{pmatrix} \qquad (3.22)$$

are the new inputs. Then

$$\ddot{\eta} = \tilde{\tau} \qquad (3.23)$$

Rewriting equations (3.19)–(3.20):

$$m\ddot{x} = -u \sin\theta \qquad (3.24)$$
$$m\ddot{y} = u \cos\theta \sin\phi \qquad (3.25)$$
$$m\ddot{z} = u \cos\theta \cos\phi - mg \qquad (3.26)$$
$$\ddot{\psi} = \tilde{\tau}_\psi \qquad (3.27)$$
$$\ddot{\theta} = \tilde{\tau}_\theta \qquad (3.28)$$
$$\ddot{\phi} = \tilde{\tau}_\phi \qquad (3.29)$$

where x and y are the coordinates in the horizontal plane, and z is the vertical position (see Figure 3.5). ψ is the yaw angle around the z-axis, θ is the pitch angle around the (new) y-axis, and ϕ is the roll angle around the (new) x-axis. The control inputs u, $\tilde{\tau}_\psi$, $\tilde{\tau}_\theta$ and $\tilde{\tau}_\phi$ are the total thrust or collective input (directed out from the bottom of the aircraft) and the new angular moments (yawing moment, pitching moment and rolling moment).

3.3 Control Strategy

In this section we will develop a control strategy for stabilizing the rotorcraft at hover.

The controller synthesis procedure regulates each of the state variables in a sequence using a priority rule as follows:

We first design a control to stabilize the yaw angular displacement. We then control the roll angle ϕ and the y-displacement using a controller designed for a PVTOL.

Finally the pitch angle θ and the x-displacement are controlled using again a strategy designed for the PVTOL.

The proposed control strategy is relatively simple. This property is important when performing real-time applications. Furthermore, the four control inputs can independently operate in either manual or automatic mode (Figure 3.6). For flight safety reasons this feature is particularly important when implementing the control strategy.

Roughly speaking each of the control inputs can be used to control one or two degrees of freedom as follows: The control input u is essentially used to make the altitude reach a desired value. The control input $\tilde{\tau}_\psi$ is used to set the yaw displacement to zero. $\tilde{\tau}_\theta$ is used to control the pitch angle and the horizontal movement in the x-axis. Similarly $\tilde{\tau}_\phi$ is used to control the roll angle and the horizontal displacement in the y-axis (see Table 3.1).

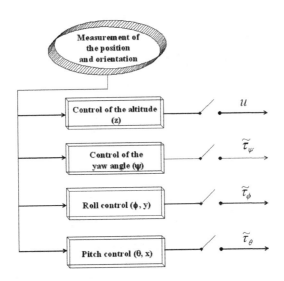

Fig. 3.6. Manual/Automatic switch diagram.

Phase	Name	Description
1	Control altitude	u is used to make the altitude reach a desired value
2	Yaw control	$\tilde{\tau}_\psi$ is used to set the yaw displacement to zero
3	Roll control	$\tilde{\tau}_\phi$ is used to control the roll ϕ and the horizontal movement in the y-axis
4	Pitch control	$\tilde{\tau}_\theta$ is used to control the pitch θ and the horizontal movement in the x-axis

Table 3.1. Control strategy sequence.

3.3.1 Altitude and Yaw Control

The control of the vertical position can be obtained by using the following control input.

$$u = (r_1 + mg)\frac{1}{c_\theta c_\phi} \tag{3.30}$$

where

$$r_1 = -a_{z_1}\dot{z} - a_{z_2}(z - z_d) \tag{3.31}$$

where a_{z_1}, a_{z_2} are positive constants and z_d is the desired altitude. The yaw angular position can be controlled by applying

$$\tilde{\tau}_\psi = -a_{\psi_1}\dot{\psi} - a_{\psi_2}(\psi - \psi_d) \tag{3.32}$$

Indeed, introducing (3.30)–(3.32) into (3.24)–(3.27) and provided that $c_\theta c_\phi \neq 0$, we obtain

$$m\ddot{x} = -(r_1 + mg)\frac{\tan\theta}{\cos\phi} \tag{3.33}$$

$$m\ddot{y} = (r_1 + mg)\tan\phi \tag{3.34}$$

$$\ddot{z} = \frac{1}{m}(-a_{z_1}\dot{z} - a_{z_2}(z - z_d)) \tag{3.35}$$

$$\ddot{\psi} = -a_{\psi_1}\dot{\psi} - a_{\psi_2}(\psi - \psi_d) \tag{3.36}$$

The control parameters a_{ψ_1}, a_{ψ_2}, a_{z_1}, a_{z_2} should be carefully chosen to ensure a stable well-damped response in the vertical and yaw axes.

From equations (3.35) and (3.36) it follows that $\psi \to \psi_d$ and $z \to z_d$.

3.3.2 Roll Control (ϕ, y)

Note that from (3.31) and (3.35) $r_1 \to 0$. For a time T large enough, r_1 and ψ are arbitrarily small therefore (3.33) and (3.34) reduce to

$$\ddot{x} = -g\frac{\tan\theta}{\cos\phi} \tag{3.37}$$

$$\ddot{y} = g\tan\phi \tag{3.38}$$

We will first consider the subsystem given by (3.29) and (3.38). We will implement a nonlinear control based on nested saturations. This type of control allows in the limit a guarantee of arbitrary bounds for ϕ, $\dot{\phi}$, y and $\dot{y}$. To further simplify the analysis we will impose a very small upper bound on $|\phi|$ in such a way that the difference $\tan(\phi) - \phi$ is arbitrarily small. Therefore, the subsystem (3.29)–(3.38) reduces to

$$\ddot{y} = g\phi \tag{3.39}$$

$$\ddot{\phi} = \tilde{\tau}_\phi \tag{3.40}$$

which represents four integrators in cascade. Then, we propose

$$\tilde{\tau}_\phi = -\sigma_{\phi_1}(\dot{\phi} + \sigma_{\phi_2}(\zeta_{\phi_1})) \tag{3.41}$$

where $\sigma_i(s)$ is a saturation function such that $|\sigma_i(s)| \leq M_i$ for $i = 1, ..., 4$ (see Figure 3.7) and ζ_{ϕ_1} will be defined later to ensure global stability.

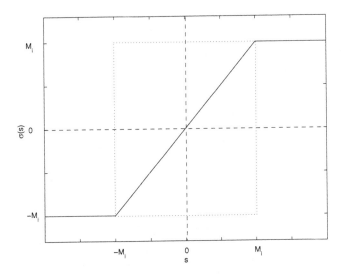

Fig. 3.7. Saturation function.

We propose the following positive function

$$V = \frac{1}{2}\dot{\phi}^2 \tag{3.42}$$

Differentiating V with respect to time, we obtain

$$\dot{V} = \dot{\phi}\ddot{\phi} \tag{3.43}$$

and from equations (3.40) and (3.41) we have

$$\dot{V} = -\dot{\phi}\sigma_{\phi_1}(\dot{\phi} + \sigma_{\phi_2}(\zeta_{\phi_1})) \tag{3.44}$$

Note that if $|\dot{\phi}| > M_{\phi_2}$ then $\dot{V} < 0$, i.e. $\exists\ T_1$ such that $|\dot{\phi}| \leq M_{\phi_2}$ for $t > T_1$.

We define

$$\nu_1 \equiv \phi + \dot{\phi} \tag{3.45}$$

Differentiating (3.45)

$$\dot{\nu}_1 = \dot{\phi} + \ddot{\phi} \tag{3.46}$$

$$= \dot{\phi} - \sigma_{\phi_1}(\dot{\phi} + \sigma_{\phi_2}(\zeta_{\phi_1})) \tag{3.47}$$

Let us choose

$$M_{\phi_1} \geq 2M_{\phi_2} \tag{3.48}$$

From the definition of $\sigma(s)$ we can see that $|\sigma_i(s)| \leq M_i$. This implies that in a finite time, $\exists\ T_1$ such that $|\dot{\phi}| \leq M_{\phi_2}$ for $t \geq T_1$. Therefore, for $t \geq T_1$, $|\dot{\phi} + \sigma_{\phi_2}(\zeta_{\phi_1})| \leq 2M_{\phi_2}$. It then follows that, $\forall\ t \geq T_1$

$$\sigma_{\phi_1}(\dot{\phi} + \sigma_{\phi_2}(\zeta_{\phi_1})) = \dot{\phi} + \sigma_{\phi_2}(\zeta_{\phi_1}) \tag{3.49}$$

Using (3.47) and (3.49), we get

$$\dot{\nu}_1 = -\sigma_{\phi_2}(\zeta_{\phi_1}) \tag{3.50}$$

Let us define

$$\zeta_{\phi_1} \equiv \nu_1 + \sigma_{\phi_3}(\zeta_{\phi_2}) \tag{3.51}$$

Introducing the above in (3.50) it follows that

$$\dot{\nu}_1 = -\sigma_{\phi_2}(\nu_1 + \sigma_{\phi_3}(\zeta_{\phi_2})) \tag{3.52}$$

The upper bounds are assumed to satisfy

$$M_{\phi_2} \geq 2M_{\phi_3} \tag{3.53}$$

This implies that $\exists\ T_2$ such that $|\nu_1| \leq M_{\phi_3}$ for $t \geq T_2$. From equation (3.45) we can see that $\forall\ t \geq T_2$, $|\phi| \leq M_{\phi_3}$. M_{ϕ_3} should be chosen small enough such that $\tan(\phi) \approx \phi$.

From (3.52) and (3.53), we have that for t$\geq T_2$, $|\nu_1 + \sigma_{\phi_3}(\zeta_{\phi_2})| \leq 2M_{\phi_3}$. It then follows that, $\forall\, t \geq T_2$

$$\sigma_{\phi_2}(\nu_1 + \sigma_{\phi_3}(\zeta_{\phi_2})) = \nu_1 + \sigma_{\phi_3}(\zeta_{\phi_2}) \tag{3.54}$$

Introducing the function

$$\nu_2 \equiv \nu_1 + \phi + \frac{\dot{y}}{g} \tag{3.55}$$

then

$$\dot{\nu}_2 = \dot{\nu}_1 + \dot{\phi} + \frac{\ddot{y}}{g} \tag{3.56}$$

Using (3.39), (3.45), (3.52) and (3.54) in (3.56), we obtain

$$\dot{\nu}_2 = -\sigma_{\phi_3}(\zeta_{\phi_2}) \tag{3.57}$$

Now, define ζ_{ϕ_2} as

$$\zeta_{\phi_2} \equiv \nu_2 + \sigma_{\phi_4}(\zeta_{\phi_3}) \tag{3.58}$$

Let us rewrite (3.57) as

$$\dot{\nu}_2 = -\sigma_{\phi_3}(\nu_2 + \sigma_{\phi_4}(\zeta_{\phi_3})) \tag{3.59}$$

We chose

$$M_{\phi_3} \geq 2M_{\phi_4} \tag{3.60}$$

We then have that in a finite time, $\exists\, T_3$ such that $|\nu_2| \leq M_{\phi_4}$ for $t \geq T_3$, this implies from (3.55) that $\dot{y}$ is bounded.

For t$\geq T_3$, $|\nu_2 + \sigma_{\phi_4}(\zeta_{\phi_3})| \leq 2M_{\phi_4}$. It then follows that, $\forall\, t \geq T_3$

$$\sigma_{\phi_3}(\nu_2 + \sigma_{\phi_4}(\zeta_{\phi_3})) = \nu_2 + \sigma_{\phi_4}(\zeta_{\phi_3}) \tag{3.61}$$

Defining

$$\nu_3 \equiv \nu_2 + 2\frac{\dot{y}}{g} + \phi + \frac{y}{g} \tag{3.62}$$

then

$$\dot{\nu}_3 = \dot{\nu}_2 + 2\frac{\ddot{y}}{g} + \dot{\phi} + \frac{\dot{y}}{g} \tag{3.63}$$

Finally using (3.39), (3.54), (3.59) and (3.61) in (3.63), we obtain

$$\dot{\nu}_3 = -\sigma_{\phi_4}(\zeta_{\phi_3}) \tag{3.64}$$

We propose ζ_{ϕ_3} in the following form

$$\zeta_{\phi_3} \equiv \nu_3 \tag{3.65}$$

then

$$\dot{\nu}_3 = -\sigma_{\phi_4}(\nu_3) \tag{3.66}$$

and this implies that $\nu_3 \to 0$. From (3.59) it follows that $\nu_2 \to 0$ and from equation (3.58) $\zeta_{\phi_2} \to 0$. From (3.52) $\nu_1 \to 0$ then from (3.51) $\zeta_{\phi_1} \to 0$.

We can see from equation (3.44) that $\dot{\phi} \to 0$. From equation (3.45) we get $\phi \to 0$. From (3.55) $\dot{y} \to 0$ and finally from (3.62) $y \to 0$.

Using (3.45), (3.51), (3.55), (3.58), (3.62) and (3.65), we can rewrite equation (3.41) as

$$\tilde{\tau}_\phi = -\sigma_{\phi_1}(\dot{\phi} + \sigma_{\phi_2}(\phi + \dot{\phi} + \sigma_{\phi_3}(2\phi + \dot{\phi} + \frac{\dot{y}}{g} + \sigma_{\phi_4}(\dot{\phi} + 3\phi + 3\frac{\dot{y}}{g} + \frac{y}{g})))) \tag{3.67}$$

3.3.3 Pitch Control (θ, x)

From equations (3.39) and (3.67) we obtain $\phi \to 0$, then (3.37) gives

$$\ddot{x} = -g \tan \theta \tag{3.68}$$

Finally we take the subsystem

$$\ddot{x} = -g \tan \theta \tag{3.69}$$
$$\ddot{\theta} = \tilde{\tau}_\theta \tag{3.70}$$

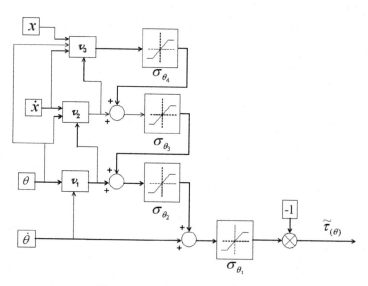

Fig. 3.8. Pitch control scheme.

As before, we assume that the control strategy will insure a very small bound on $|\theta|$ in such a way that $\tan(\theta) \approx \theta$. Therefore (3.69) reduces to

$$\ddot{x} = -g\theta \qquad (3.71)$$

Using a procedure similar to the one proposed for the roll control we obtain

$$\tilde{\tau}_\theta = -\sigma_{\theta_1}(\dot{\theta} + \sigma_{\theta_2}(\theta + \dot{\theta} + \sigma_{\theta_3}(2\theta + \dot{\theta} - \frac{\dot{x}}{g} + \sigma_{\theta_4}(\dot{\theta} + 3\theta - 3\frac{\dot{x}}{g} - \frac{x}{g}))))$$

$$(3.72)$$

Figure 3.8 represents the pitch control scheme.

3.4 Experimental Results

In this section we present the real-time experimental results obtained when applying the controller proposed in the previous section to a four-rotor mini-rotorcraft (see Figure 3.9). We will first describe the architecture of the platform and then explain how the controller parameters were tuned to ensure satisfactorily the tasks of take-off, hover and landing.

3.4.1 Platform Description

The flying machine we have used is a mini-rotorcraft having four rotors manufactured by Draganfly Innovations Inc. (*http://www.rctoys.com*). The physical characteristics of this rotorcraft are given in Table 3.2.

Weight (not including batteries)	320 g
Batteries weight	200 g
Maximum length	74 cm
Blades diameter	29 cm
Height	11 cm
Distance between the motor and the C.G.	20.5 cm
Motor reduction rate	1:6

Table 3.2. Rotorcraft physical characteristics.

The four control signals are transmitted by a Futaba Skysport 4 radio. The control signals are throttle control input u, pitch control input $\tilde{\tau}_\theta$, roll control input $\tilde{\tau}_\phi$ and yaw control input $\tilde{\tau}_\psi$. These control signals are constrained in the radio to satisfy

$$\begin{array}{l} 0.66 \text{ V} < u < 4.70 \text{ V} \\ 1.23 \text{ V} < \tilde{\tau}_\theta < 4.16 \text{ V} \\ 0.73 \text{ V} < \tilde{\tau}_\phi < 4.50 \text{ V} \\ 0.40 \text{ V} < \tilde{\tau}_\psi < 4.16 \text{ V} \end{array} \qquad (3.73)$$

Fig. 3.9. The Draganfly III.

The radio and the PC (INTEL Pentium III) are connected using data acquisition cards (ADVANTECH PCL-818HG and PCL-726). The connection in the radio is directly made to the joystick potentiometers for the collective, yaw, pitch and roll controls. In order to simplify the tuning of the controller and for flight security reasons, we have introduced several switches in the PC–radio interface so that each control input can operate either in manual mode or in automatic control mode. Therefore we select the control inputs that are handled manually by the pilot while the other control inputs are provided by the computer. In the first stage only the yaw control input is controlled by the computer. After obtaining good performance in the yaw displacement we start dealing with the automatic control of the roll. In the third stage only the throttle is handled by the pilot while in the last stage all the modes are controlled by the computer.

The rotorcraft evolves freely in a 3D space without any flying stand. To measure the position (x, y, z) and orientation (ψ, θ, ϕ) of the rotorcraft we use the 3D tracker system (POLHEMUS) [46]. The Polhemus is connected via RS232 to the PC (see Figure 3.10). This type of sensor is very sensitive to electromagnetic noise and we had to install it as far as possible from the electric motors and their drivers.

The Draganfly III has three onboard gyros that help stabilize the minirotorcraft. It would be impossible for a pilot to stabilize this flying machine without gyro stabilization. It would be a very difficult task to deactivate the gyros and we think that since they work so well, it would not be reasonable to disconnect them. Thus, these three angular velocity feedbacks provided by the on-board gyros are considered as an in-built inner control loop. Notice that this gyro stabilization is not enough for performing hover autonomously.

The use of the Polhemus sensor allows the achievement of practical experiments on the laboratory. An actual UAV (Unmanned Autonomous Vehicle) would require to use on-board inertial measurements to be able to fly outdoors.

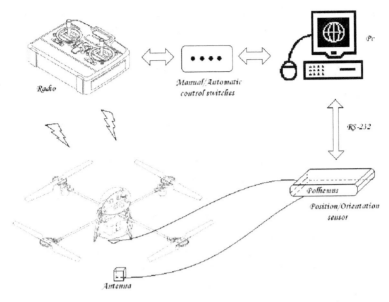

Fig. 3.10. Real-time computer architecture of the platform.

3.4.2 Controller Parameter Tuning

The computation of the control input requires knowledge of various angular and linear velocities. The sensor at our disposal only measures position and orientation. We have thus computed estimates of the angular and linear velocities by using the approximation

$$\dot{q}_t \approx \frac{q_t - q_{t-T}}{T} \tag{3.74}$$

where q is a given variable and T is the sampling period. In our experiment $T = 71$ ms due to limitations imposed by the measuring device. In order to obtain a good estimate of the angular and linear velocities and avoid abrupt changes in these signals we have introduced numerical filters.

The controller parameters are selected using the following procedure. The yaw controller parameters are first tuned while the other modes are handled by the pilot. The yaw controller is basically a PD controller. The parameters are selected to obtain a short settling time without introducing small oscillations in the yaw displacement. The yaw control input should also satisfy the constraints (3.73).

The parameters of the roll control input are carried out while the throttle and pitch controls are in manual mode. The parameters of the roll control are adjusted in the following sequence.

We first select the gain concerning roll angular velocity $\dot{\phi}$. Due to the on-board gyros, this gain is relatively small. We next select the controller gain concerning the roll displacement ϕ. We wish the roll error to converge to zero fast but without undesirable oscillations. The controller gain concerning $\dot{y}$ and the amplitude of the saturation function are selected in such a way that the mini-aircraft reduces its speed in the y-axis fast enough. To complete the tuning of the roll control parameters we choose the gains concerning the y displacement to obtain a satisfactory performance. The parameters of the pitch control are selected similarly.

Finally we tune the parameters of the throttle control to obtain a desired altitude. One of the controller parameters is used to compensate the gravity force which is estimated off-line using experimental data.

Notice that since this mini-rotorcraft has soft blades, the tuning of the parameters can be done while holding the rotorcraft in the hand and wearing eye-protection glasses. This can certainly not be done with larger flying machines and therefore more simulation developments have to be performed before actually applying the controller to the real system. In our case simulations were actually not required.

3.4.3 Experiment

The control objective is to make the mini-rotorcraft hover at an altitude of 30 cm, i.e. we wish to reach the position $(x, y, z) = (0, 0, 30)$ cm while $(\psi, \theta, \phi) = (0, 0, 0)$ °. We also make the rotorcraft follow a simple horizontal trajectory. The gain values used for the control law are given in Table 3.3.

Phase	Control parameter	Value
1.- Altitude	a_{z1}	0.001
	a_{z2}	0.002
2.- Yaw control	$a_{\psi 1}$	2.374
	$a_{\psi 2}$	0.08
3.- Roll control	$M_{\phi 1}$	2
	$M_{\phi 2}$	1
	$M_{\phi 3}$	0.2
	$M_{\phi 4}$	0.1
4.- Pitch control	$M_{\theta 1}$	2
	$M_{\theta 2}$	1
	$M_{\theta 3}$	0.2
	$M_{\theta 4}$	0.1
1-4	T	$\frac{1}{14} s$

Table 3.3. Gain values used in the control law.

Figures 3.11−3.14 show the performance of the controller when applied to the rotorcraft. Figure 3.12 shows the four-rotor rotorcraft hovering autonomously. Videos of the experiments can be seen at the following address: *http://www.hds.utc.fr/∼castillo/4r_fr.html*

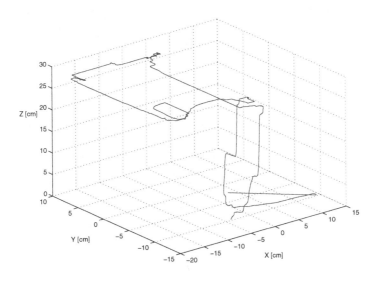

Fig. 3.11. 3D motion of the rotorcraft.

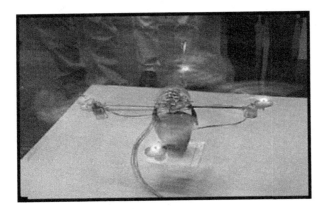

Fig. 3.12. The four-rotor rotorcraft hovering autonomously.

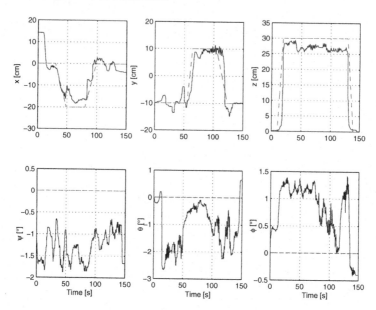

Fig. 3.13. Position (x, y, z) and orientation (ψ, θ, ϕ) of the quad-rotor rotorcraft. The dotted lines represent the desired trajectories.

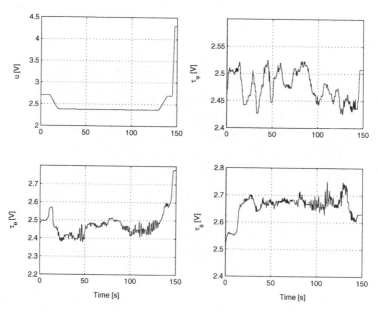

Fig. 3.14. Control inputs (throttle input, yaw, pitch and roll).

3.5 Conclusion

We have proposed a stabilization control algorithm for a mini-rotorcraft having four rotors. The dynamic model of the rotorcraft was obtained via a Lagrange approach and the proposed control algorithm is based on nested saturations.

The proposed strategy has been successfully applied to the rotorcraft, and the experimental results have shown that the controller performs satisfactorily.

To the best of our knowledge, this is the first successful real-time control applied to a four-rotor rotorcraft.

4

Robust Prediction-based Control for Unstable Delay Systems

4.1 Introduction

The area of control of delay systems has attracted the attention of many researchers in the past few years [48], [123], [189]. This is motivated by the fact that delays are responsible for instabilities in closed-loop control systems. Delays appear due to transport phenomena, computation of the control input, time-consuming information processing in measurement devices, etc.. A fundamental algorithm based on state-prediction control was proposed by [98]. In [54] a continuous-time state-predictive control system is presented. Robust stability is proved for uncertainties in the gain and the delay of the plant. [20] present an algebraic formalism using the Bezout identity, for assigning, by feedback, a finite spectrum. Within this framework [93] present the conditions for a compensator to be realizable with internal stability. Stability of delay systems based on passivity is studied in [124]. Other approaches for the control of systems with delay are available as the Smith predictor [154], [130], [106], [107], [87], and its many improved schemes generically named process-model control schemes [183], finite spectrum assignment techniques [98], reduction achieved through transformations and algebraic approaches such as those in [111], [82], [20].

A close analysis of these methods shows that they all use, in an explicit or implicit manner, prediction of the state in order to achieve control of the system. A common drawback, linked to the internal instability of the prediction, is that they fail to stabilize unstable systems.

The problems arising in the implementation of continuous-time distributed-delay control laws have been studied in [182]. These delays appear in control laws allowing finite spectrum assignment for delay systems, but cannot be exactly implemented on a digital computer. [182] shows in simulations that for a simple but unstable system the closed-loop system can be stable or unstable depending on the numerical integration approximation method that is used in the implementation.

On the other hand if both the delay system and the distributed-delay controller are discretized, then the discretized closed-loop system remains stable. Furthermore, in the ideal case, the resulting closed-loop behaviour is the same as if the original plant has no delay. This result holds if the time elapsed between sampling instants is constant and the delay is a multiple of the sampling period. In this chapter we address the problem of robustness with respect to the variations in the sampling instants, in the delay and the delay/sampling period ratio.

The well-known pole-placement controller proposed by [98] requires the computation of an integral used to predict the state. In the ideal case this control scheme leads to a finite pole-placement. However, arbitrary small errors in the computation of the integral term produce instability, as shown in [105] and [182]. This can also be explained by the fact that in the ideal case the closed-loop behaviour is governed by a finite polynomial while in the presence of small errors, the closed-loop behaviour is given by a quasi-polynomial having an infinite number of roots. Since in practice we normally use a computer to implement the control law [7], it is justified to study whether instabilities can also appear in discrete-time pole-placement control algorithms. Note that small variations in the sampling period may be such that the closed-loop behaviour will be described by a quasi-polynomial in the complex variable z. The zero location of quasi-polynomials is known to be very sensitive to small changes in the polynomial parameters and can easily move from the stable region to the unstable region. Therefore it is important to prove robustness also with respect to small variations of the sampling period. To our knowledge, this type of robustness has not been studied in the literature for discrete-time systems.

In this chapter we present the stability analysis of a hybrid control scheme, i.e. when the system representation is given in continuous-time while the controller is expressed in discrete-time [95]. The controller is basically a discrete-time state-feedback control in which the actual state is replaced by the prediction of the state. We present a stability proof based on Lyapunov analysis of the hybrid closed-loop system. Convergence of the state to the origin is ensured regardless of whether the original system is stable or not. The stability is established in spite of uncertainties in the knowledge of the plant parameters and the delay. Robustness is also proved with respect to small variations of the time elapsed between sampling instants.

The proposed prediction-based controller has been tested in a real-time application to control the yaw angular displacement of a four-rotor mini-helicopter [94]. The experimental validation of the proposed algorithm has been developed on a novel real-time system, MaRTE OS, which allows the implementation of minimum real-time systems according to standard POSIX.13 of the IEEE [140].

The chapter is organized as follows: the discrete-time representation of the system including uncertainties is presented in Section 4.2. Section 4.3 is devoted to presenting the state predictor. The prediction-based state-feedback controller is given in Section 4.4. Section 4.5 presents the stability analysis of the hybrid closed-loop system. In Section 4.6 we show the experimental results and finally, the conclusions are given in Section 4.7.

4.2 Problem Formulation

Let us consider the following continuous-time state space representation of a system with input delay

$$\dot{x}(t) = A_c x(t) + B_c u(t - h(t)) \tag{4.1}$$

where the nominal plant parameter matrices are $A_c \in \Re^{n \times n}$, $B_c \in \Re^{n \times m}$ and $h(t)$ is the time-varying plant delay. Usually, in the discrete-time framework, the sampling time instant t_k is defined as $t_k = kT$ where T is the sampling period and k is an integer. However, since we wish to prove robustness of the control scheme with respect to the time elapsed between sampling time instants, we will not define t_k as a multiple of T. We will rather define t_k as the k-th sampling instant such that

$$t_{k+1} - t_k = T + \varepsilon_k \tag{4.2}$$

where T is the ideal sampling period and ε_k is a small variation of the time elapsed between sampling instants. Furthermore, we will assume that T and h satisfy

$$h(t) = dT + \epsilon(t) \tag{4.3}$$

where d is an integer and $\epsilon(t)$ is a small uncertainty in the knowledge of the delay $h(t)$. Both variations ε_k and $\epsilon(t)$ can be positive or negative but they should be bounded as follows

$$|\varepsilon_k| \leq \bar{\varepsilon} \ll T$$
$$|\epsilon(t)| \leq \bar{\epsilon} \ll T$$

We will use the notation $x_k = x(t_k)$. From (4.1) we obtain the following time response equation

$$x_{k+1} = A_1 x_k + \int_{t_k}^{t_{k+1}} e^{A_c(t_{k+1}-\tau)} B_c u(\tau - h(\tau)) d\tau \tag{4.4}$$

where

$$A_1 = e^{A_c(t_{k+1}-t_k)} = e^{A_c(T+\varepsilon_k)} \tag{4.5}$$

We will define A as

$$A = e^{A_c T} \tag{4.6}$$

and Δ_4 such that

$$A_1 = A + \Delta_4 \tag{4.7}$$

Since we are interested in implementing the control law in a computer, we will assume that the input u is constant between sampling instants, i.e. $u(t) = u_k \ \forall \ t \in [t_k, t_{k+1})$.

The following lemma shows how the recursive equation for x_k is modified due to the uncertainties in the plant parameters A_c and B_c, the delay $h(t)$ and the ideal sampling period T.

Lemma 4.1. *The recursive equation for system (4.1) for time sampling instants defined in (4.2) and the delay in (4.3) is given by*

$$x_{k+1} = Ax_k + Bu_{k-d} + \Delta f_k \tag{4.8}$$

where

$$f_k = \left[u_{k-d-1}^T, u_{k-d}^T, u_{k-d+1}^T, x_k^T \right]^T \tag{4.9}$$

and Δ is a matrix which is bounded by $\bar{\varepsilon}$ and $\bar{\epsilon}$. Therefore Δ converges to zero as $\bar{\varepsilon}$ and $\bar{\epsilon}$ converge to zero.

Proof. Let us obtain an expression for $h(\tau)$ in (4.4). From (4.2) we get

$$\sum_{i=k-d}^{k-1} (t_{i+1} - t_i - T - \varepsilon_i) = 0 \tag{4.10}$$

or

$$dT = t_k - t_{k-d} - \mu_k \tag{4.11}$$

where

$$\mu_k = \sum_{i=k-d}^{k-1} \varepsilon_i \tag{4.12}$$

Introducing (4.11) into (4.3) we obtain

$$\begin{aligned} h(\tau) &= dT + \epsilon(\tau) \\ &= t_k - t_{k-d} - \mu_k + \epsilon(\tau) \end{aligned} \tag{4.13}$$

Therefore, the last term in (4.4) can be rewritten as

$$M(k) \triangleq \int_{t_k}^{t_{k+1}} e^{A_c(t_{k+1}-\tau)} B_c u(\tau - h(\tau)) d\tau$$

$$= \int_{t_k}^{t_k+T} e^{A_c(t_{k+1}-\tau)} B_c u(\tau - h(\tau)) d\tau + \delta_1$$

$$= \int_{t_k}^{t_k+T} e^{A_c(t_k+T-\tau)} B_c u(\tau - h(\tau)) d\tau + \delta_1 + \delta_2 \tag{4.14}$$

$$= \int_{t_k}^{t_k+T} e^{A_c(t_k+T-\tau)} B_c u(\tau - t_k + t_{k-d}) d\tau + \delta_1 + \delta_2 + \delta_3$$

$$= Bu_{k-d} + \delta_1 + \delta_2 + \delta_3 + \delta_4$$

where (see (4.2), (4.13))

$$\delta_1 = \int_{t_k+T}^{t_k+T+\varepsilon_k} e^{A_c(t_{k+1}-\tau)} B_c u(\tau - h(\tau)) d\tau \qquad (4.15)$$

$$\delta_2 = \int_{t_k}^{t_k+T} \left[e^{A_c(t_k+T+\varepsilon_k-\tau)} - e^{A_c(t_k+T-\tau)} \right] B_c u(\tau - h(\tau)) d\tau$$

$$= \int_{t_k}^{t_k+T} e^{A_c(t_k+T-\tau)} \left[e^{A_c\varepsilon_k} - I \right] B_c u(\tau - h(\tau)) d\tau \qquad (4.16)$$

$$\delta_3 = \int_{t_k}^{t_k+T} e^{A_c(t_k+T-\tau)} B_c [u(\tau - t_k + t_{k-d} + \mu_k - \epsilon(\tau))$$

$$-u(\tau - t_k + t_{k-d})] d\tau \qquad (4.17)$$

$$\delta_4 = \int_{t_k}^{t_k+T} e^{A_c(t_k+T-\tau)} B_c \left[u(\tau - t_k + t_{k-d}) - u(t_{k-d}) \right] d\tau$$

$$= \int_{t_k+T+\varepsilon_{k-d}}^{t_k+T} e^{A_c(t_k+T-\tau)} B_c [u(\tau - t_k + t_{k-d}) - u(t_{k-d})] d\tau \quad (4.18)$$

(In the above we have used the fact that $u(\tau - t_k + t_{k-d}) = u(t_{k-d})$ for $\tau \in [t_k, t_k + T + \varepsilon_{k-d})$.)

$$B = \int_{t_k}^{t_k+T} e^{A_c(t_k+T-\tau)} B_c d\tau$$

$$= \int_0^T e^{A_c(T-s)} B_c ds \qquad (4.19)$$

Define

$$\nu(\tau, k) \triangleq u(\tau - t_k + t_{k-d} + \mu_k - \epsilon(\tau)) - u(\tau - t_k + t_{k-d}) \qquad (4.20)$$

Assume that λ is the upper bound for $\mu_k - \epsilon(\tau)$, i.e.

$$|\mu_k - \epsilon(\tau)| \leq d\bar{\varepsilon} + \bar{\varepsilon} \triangleq \lambda \ll T \qquad (4.21)$$

Then δ_3 in (4.17) can be written as

$$\delta_3 = \delta_5 + \delta_6 + \delta_7 \qquad (4.22)$$

where

$$\delta_5 = \int_{t_k}^{t_k+\lambda} e^{A_c(t_k+T-\tau)} B_c \nu(\tau, k) d\tau \qquad (4.23)$$

$$\delta_6 = \int_{t_k+\lambda}^{t_k'} e^{A_c(t_k+T-\tau)} B_c \nu(\tau, k) d\tau \tag{4.24}$$

$$\delta_7 = \int_{t_k'}^{t_k+T} e^{A_c(t_k+T-\tau)} B_c \nu(\tau, k) d\tau \tag{4.25}$$

where (see (4.2))

$$t_k' = t_k + T + \varepsilon_{k-d} - \lambda$$

$$= t_k + t_{k-d+1} - t_{k-d} - \lambda \tag{4.26}$$

Notice from (4.20) that $\nu(\tau, k) = 0$ for $\tau \in \left[t_k + \lambda, t_k'\right]$. Thus, $\delta_6 = 0$.

From (4.15), using a change of variable, δ_1 can also be written as

$$\delta_1 = \int_{-\varepsilon_k}^{0} e^{-A_c s} B_c u(s - h(s + t_k + T)) ds \tag{4.27}$$

In general, depending on the values of the uncertainties ε_k and $\epsilon(t)$, $u(s - h(s + t_k + T))$ in the interval $s \in [0, \varepsilon_k]$ is equal to either u_{k-d-1}, u_{k-d} or u_{k-d+1}. Let us define those corresponding time intervals as Ω_{11}, Ω_{12} and Ω_{13} respectively. Therefore δ_1 can be expressed as

$$\delta_1 = \int_{\Omega_{11}} e^{-A_c s} B_c ds\, u_{k-d-1} + \int_{\Omega_{12}} e^{-A_c s} B_c ds\, u_{k-d} + \int_{\Omega_{13}} e^{-A_c s} B_c ds\, u_{k-d+1} \tag{4.28}$$

or, with obvious notation

$$\delta_1 = \Gamma_{11}\, u_{k-d-1} + \Gamma_{12} u_{k-d} + \Gamma_{13} u_{k-d+1} \tag{4.29}$$

Notice that the length of the intervals Ω_{11}, Ω_{12} and Ω_{13} is bounded by $\bar{\varepsilon}$. Similarly we can define

$$\delta_j = \Gamma_{j1}\, u_{k-d-1} + \Gamma_{j2} u_{k-d} + \Gamma_{j3} u_{k-d+1} \tag{4.30}$$

for $j = 4, 5, 7$. Recall that $\delta_6 = 0$. Notice that the length of the intervals Ω_{jl} for $l = 1, 2, 3$ is bounded by $\bar{\varepsilon}$ for $j = 4$ (see (4.18)), $\lambda = d\, \bar{\varepsilon} + \bar{\epsilon}$ for $j = 5$ (see (4.23)), and $(d + 1)\, \bar{\varepsilon} + \bar{\epsilon}$ for $j = 7$ (see (4.25)). From (4.16), δ_2 can be rewritten as

$$\delta_2 = \int_{\Omega_{21}} e^{-A_c s} \left[e^{A_c \varepsilon_k} - I \right] B_c ds\, u_{k-d-1}$$

$$+ \int_{\Omega_{22}} e^{-A_c s} \left[e^{A_c \varepsilon_k} - I \right] B_c ds\, u_{k-d} \tag{4.31}$$

$$+ \int_{\Omega_{23}} e^{-A_c s} \left[e^{A_c \varepsilon_k} - I \right] B_c ds\, u_{k-d+1}$$

or with obvious notation

$$\delta_2 = \Gamma_{21}\, u_{k-d-1} + \Gamma_{22} u_{k-d} + \Gamma_{23} u_{k-d+1} \tag{4.32}$$

The length of the intervals Ω_{2l} for $l = 1, 2, 3$ is bounded by T. In view of the above we can compute the following upper bounds for $l = 1, 2, 3$

$$\|\Gamma_{1l}\| \leq \bar{\varepsilon} \max_{-\bar{\varepsilon} \leq s \leq \bar{\varepsilon}} \left\| e^{-A_c s} B_c \right\|$$

$$\|\Gamma_{2l}\| \leq T \max_{0 \leq s \leq T} \left\| e^{A_c s} \right\| \|B_c\| \max_{-\bar{\varepsilon} \leq s \leq \bar{\varepsilon}} \left\| e^{A_c s} - I \right\|$$

$$\|\Gamma_{4l}\| \leq \bar{\varepsilon} \max_{-\bar{\varepsilon} \leq s \leq \bar{\varepsilon}} \left\| e^{-A_c s} B_c \right\| \tag{4.33}$$

$$\|\Gamma_{5l}\| \leq (d\bar{\varepsilon} + \bar{\epsilon}) \max_{-\bar{\varepsilon} \leq s \leq \bar{\varepsilon}} \left\| e^{-A_c s} B_c \right\|$$

$$\|\Gamma_{7l}\| \leq ((d+1)\bar{\varepsilon} + \bar{\epsilon}) \max_{-\bar{\varepsilon} \leq s \leq \bar{\varepsilon}} \left\| e^{-A_c s} B_c \right\|$$

Note also that

$$\left\| e^{A_c s} - I \right\| = \left\| \sum_{i=0}^{\infty} \frac{(A_c s)^i}{i!} - I \right\|$$

$$= \left\| \sum_{i=1}^{\infty} \frac{(A_c s)^i}{i!} \right\|$$

$$= \left\| A_c s \sum_{i=1}^{\infty} \frac{(A_c s)^{i-1}}{i!} \right\|$$

$$= \left\| A_c s \sum_{j=0}^{\infty} \frac{(A_c s)^j}{(j+1)!} \right\|$$

$$\leq \left\| A_c s \sum_{j=0}^{\infty} \frac{(A_c s)^j}{j!} \right\|$$

$$\leq \|A_c s\| \left\| e^{A_c s} \right\| \tag{4.34}$$

Therefore, $\|\Gamma_{2l}\|$ in (4.33) is also clearly bounded by $\bar{\varepsilon}$. Finally $M(k)$ in (4.14) can be expressed as in (4.36) with

$$\Delta_1 = \Gamma_{11} + \Gamma_{21} + \Gamma_{41} + \Gamma_{51} + \Gamma_{71}$$

$$\Delta_2 = \Gamma_{12} + \Gamma_{22} + \Gamma_{42} + \Gamma_{52} + \Gamma_{72} \tag{4.35}$$

$$\Delta_3 = \Gamma_{13} + \Gamma_{23} + \Gamma_{43} + \Gamma_{53} + \Gamma_{73}$$

and $\|\Gamma_{jl}\|$ for $j = 1, 2, 4, 7$ and $l = 1, 2, 3$ is bounded by $\bar{\varepsilon}$ and $\bar{\epsilon}$.

Notice that depending on the values of the uncertainties, $u(\tau - h(\tau))$ in (4.14) is a function of u_{k-d-1}, u_{k-d}, and u_{k-d+1} in the interval $[t_k, t_{k+1}]$. Therefore the δ_i for $i = 1, ..., 7$ are functions of u_{k-d-1}, u_{k-d}, and u_{k-d+1}. Notice also that the integration periods in δ_1 and δ_3 through δ_7 converge to zero as $\bar{\varepsilon}$ and $\bar{\epsilon}$ converge to zero. Furthermore $e^{A_c(t_k+T+\varepsilon_k-\tau)} - e^{A_c(t_k+T-\tau)}$ in (4.16) converges to zero as $\bar{\varepsilon}$ converges to zero (see (4.34)). From the above it follows that $M(k)$ in (4.14) can be expressed as

$$M(k) = Bu_{k-d} + \Delta_1 u_{k-d-1} + \Delta_2 u_{k-d} + \Delta_3 u_{k-d+1} \qquad (4.36)$$

for some matrices $\Delta_i \in \Re^{n \times m}$ for $i = 1, 2, 3$ that converge to zero as $\bar{\varepsilon}$ and $\bar{\epsilon}$ converge to zero.

Introducing (4.36) and (4.7) into (4.4) we obtain

$$\begin{aligned} x_{k+1} &= Ax_k + Bu_{k-d} + \Delta_1 u_{k-d-1} + \Delta_2 u_{k-d} + \Delta_3 u_{k-d+1} + \Delta_4 x_k \\ &= Ax_k + Bu_{k-d} + \Delta f_k \end{aligned} \qquad (4.37)$$

and $\Delta \in \Re^{n \times s}$ and $f_k \in \Re^s$ with $s = 3m + n$ are defined as

$$\Delta = [\Delta_1, \Delta_2, \Delta_3, \Delta_4]$$
$$f_k = \left[u_{k-d-1}^T, u_{k-d}^T, u_{k-d+1}^T, x_k^T \right]^T \qquad (4.38)$$

∎

(4.8) (or (4.37)) can be viewed as a general state-space representation for discrete-time systems in which Δ takes into account uncertainties in matrices A_c and B_c, in the delay h and in the ideal sampling period T. We assume that the nominal plant parameters A_c and B_c and the ideal sampling period T are such that (A, B) is a controllable pair. To prove robustness of the control scheme we will mainly use the property that $\Delta \to 0$ as the uncertainties in A_c, B_c, h (i.e. ϵ) and T (i.e. ε) go to zero.

4.3 d-Step Ahead Prediction Scheme

In this section we will extend the ideas in [60] to compute a d-step ahead prediction of the state in the case of the linear system with uncertainties (4.37). For simplicity of notation we have dropped the subscript k from the uncertainties Δ. From (4.37) the prediction of x_{k+2} is given by

$$\begin{aligned} x_{k+2} &= A(Ax_k + Bu_{k-d} + \Delta f_k) + Bu_{k-d+1} + \Delta f_{k+1} \\ &= A^2 x_k + ABu_{k-d} + Bu_{k-d+1} + A\Delta f_k + \Delta f_{k+1} \end{aligned} \qquad (4.39)$$

Similarly we have

$$
\begin{aligned}
x_{k+3} &= A(A^2 x_k + ABu_{k-d} + Bu_{k-d+1} + A\Delta f_k + \Delta f_{k+1}) \\
&\quad + Bu_{k-d+2} + \Delta f_{k+2} \\
&= A^3 x_k + A^2 Bu_{k-d} + ABu_{k-d+1} + Bu_{k-d+2} \\
&\quad + A^2 \Delta f_k + A\Delta f_{k+1} + \Delta f_{k+2}
\end{aligned}
\tag{4.40}
$$

Extending this prediction d steps ahead we have

$$
\begin{aligned}
x_{k+d} &= A^d x_k + A^{d-1} Bu_{k-d} + \dots + ABu_{k-2} \\
&\quad + Bu_{k-1} + A^{d-1}\Delta f_k + A^{d-2}\Delta f_{k+1} \\
&\quad + \dots + \Delta f_{k+d-1}
\end{aligned}
\tag{4.41}
$$

or

$$
\begin{aligned}
x_{k+d} &= A^d x_k + A^{d-1} Bu_{k-d} + \dots + ABu_{k-2} \\
&\quad + Bu_{k-1} + \bar{\Delta}\, \bar{f}_{k+d-1}
\end{aligned}
\tag{4.42}
$$

where $\bar{\Delta}$ and $\bar{f}_{k+d-1}$ are given by

$$
\bar{\Delta} = \left[A^{d-1}\Delta, \; A^{d-2}\Delta, \dots, \Delta \right]
\tag{4.43}
$$

and

$$
\bar{f}_{k+d-1} = \left[f_k^T, \; f_{k+1}^T, \dots, \; f_{k+d-1}^T \right]^T
\tag{4.44}
$$

Define x_{k+d}^p as the prediction of the state x_{k+d} at time t_k

$$
x_{k+d}^p = A^d x_k + A^{d-1} Bu_{k-d} + \dots + Bu_{k-1}
\tag{4.45}
$$

Note that x_{k+d}^p can be computed with information available at time t_k.

4.4 Prediction-based State Feedback Control

In this section we will define a prediction-based controller following the ideas of [60]. We will prove that our controller is robust with respect to uncertainties that are small enough. Consider the control input

$$
u_k = K^T x_{k+d}^p
\tag{4.46}
$$

or using (4.45)

$$
u_k = K^T (A^d x_k + A^{d-1} Bu_{k-d} + \dots + Bu_{k-1})
\tag{4.47}
$$

From the above and (4.42) it follows that

$$
u_k = K^T (x_{k+d} - \bar{\Delta}\, \bar{f}_{k+d-1})
\tag{4.48}
$$

Introducing the above equation into (4.37) we obtain

$$x_{k+1} = (A + BK^T)x_k - BK^T \bar{\Delta} \, \bar{f}_{k-1} + \Delta f_k \qquad (4.49)$$

As will be shown next, for small parameter and delay uncertainties, the stability of the above system will be insured if $A + BK^T$ is stable and if we can show that $\bar{f}_{k-1}$ and f_k are linear combinations of the elements of the closed-loop system state

$$z_k = \left[x_k^T, ..., x_{k-d}^T, u_{k-d-1}^T, ..., u_{k-2d-1}^T \right]^T \qquad (4.50)$$

where $z_k \in \Re^l$ with $l = (d+1)(n+m)$. Recall from (4.37) and (4.38) that

$$\Delta f_k = \Delta_1 u_{k-d-1} + \Delta_2 u_{k-d} + \Delta_3 u_{k-d+1} + \Delta_4 x_k \qquad (4.51)$$

In the above equation u_{k-d-1} and x_k are clearly elements of z_k in (4.50). Using (4.47), u_{k-d} above can be expressed in terms of $x_{k-d}, u_{k-2d}, ...,$ and u_{k-d-1} which are elements of z_k. Similarly, u_{k-d+1} can be expressed in terms of $x_{k-d+1}, u_{k-2d+1}, ...,$ and u_{k-d}. As before, u_{k-d} can be expressed in terms of elements of z_k. Therefore f_k in (4.49) can be expressed as a function of the elements of z_k. Note also that we can prove similarly that f_{k-1} is a function of z_k.

From (4.43) and (4.44) we have

$$\bar{\Delta} \, \bar{f}_{k-1} = A^{d-1}\Delta f_{k-d} + A^{d-2}\Delta f_{k-d+1} + ... + \Delta f_{k-1} \qquad (4.52)$$

In view of (4.38), $f_{k-d}, f_{k-d+1}, ...,$ and f_{k-2} in the above equation, are functions of z_k in (4.50). As explained before f_{k-1} is also a function of z_k and we conclude that $\bar{f}_{k-1}$ in (4.52) is a function of z_k. Therefore the term $-BK^T \bar{\Delta} \, \bar{f}_{k-1} + \Delta f_k$ in (4.49) can be expressed as

$$-BK^T \bar{\Delta} \, \bar{f}_{k-1} + \Delta f_k = \Delta' z_k \qquad (4.53)$$

where Δ' is a matrix whose elements vanish as Δ goes to zero. From (4.48) we get

$$u_{k-d} = K^T(x_k - \bar{\Delta} \, \bar{f}_{k-1}) \qquad (4.54)$$

or

$$u_{k-d} = K^T x_k + \Delta'' z_k \qquad (4.55)$$

where $\Delta'' = K^T \bar{\Delta}$ is a matrix whose elements vanish as Δ goes to zero. From (4.49), (4.53) and (4.55), the closed-loop system can be written as

$$
\begin{pmatrix} x_{k+1} \\ x_k \\ \vdots \\ x_{k-d+1} \\ u_{k-d} \\ u_{k-d-1} \\ \vdots \\ u_{k-2d} \end{pmatrix} = \begin{bmatrix} (A+BK^T) & 0 & 0 & \cdots\cdots\cdots\cdots & 0 \\ 1 & 0 & 0 & \cdots\cdots\cdots\cdots & 0 \\ \vdots & & \ddots & \ddots & \cdots\cdots\cdots & \vdots \\ 0 & & 0 & 1 & \ddots\cdots\cdots & 0 \\ K^T & & 0 & & \ddots\cdots\cdots\cdots & 0 \\ \vdots & & & \cdots\cdots & \ddots & 1 & \ddots\cdots & \vdots \\ \vdots & & & \cdots\cdots\cdots & & \ddots & \ddots & \vdots \\ 0 & & 0 & \cdots\cdots & 0 & 1 & 0 \end{bmatrix} \begin{pmatrix} x_k \\ x_{k-1} \\ \vdots \\ x_{k-d} \\ u_{k-d-1} \\ u_{k-d-2} \\ \vdots \\ u_{k-2d-1} \end{pmatrix} + \begin{bmatrix} \Delta' \\ 0 \\ \vdots \\ 0 \\ \Delta'' \\ 0 \\ \vdots \\ 0 \end{bmatrix} z_k
$$

$$(4.56)$$

With obvious notation we rewrite the above system as

$$z_{k+1} = \bar{A} z_k + \bar{B} z_k \tag{4.57}$$

where $\bar{B} \to 0$ as $\Delta \to 0$ and $\bar{A} \in \Re^{l \times l}$, $\bar{B} \in \Re^{l \times l}$ with $l = (d+1)(n+m)$. Note that from (4.3) it follows that $d \to \infty$ as $T \to 0$. This means that as $T \to 0$ the closed-loop system in (4.56) becomes infinite dimensional. In the following section we present a stability analysis of the closed-loop system (4.56) when $T \neq 0$ i.e. when the dimension of z_k in (4.56) is finite.

4.5 Stability of the Closed-loop System

We will now prove the stability of the closed-loop system in (4.56) or (4.57) and robustness with respect to small uncertainties in A_c, B_c, h and T in the system (4.1). It can be seen from (4.56) and (4.57) that the eigenvalues of $\bar{A}$ are given by the set of the n eigenvalues of $(A + BK^T)$ and $(l - n)$ eigenvalues at the origin. If K is chosen such that $(A + BK^T)$ is a Schur matrix, then $\bar{A}$ is also a Schur matrix, i.e. $\bar{A}$ has all its eigenvalues strictly inside the unit circle [70]. It then follows that for every $Q > 0$, $\exists\, P > 0$ such that the following Lyapunov equation holds

$$\bar{A}^T P \bar{A} - P = -Q \tag{4.58}$$

Let us define the candidate Lyapunov function V_k

$$V_k = z_k^T P z_k \tag{4.59}$$

From (4.57), (4.58) and (4.59) we have

$$
\begin{aligned}
V_{k+1} &= z_{k+1}^T P z_{k+1} \\
&= (\bar{A} z_k + \bar{B} z_k)^T P (\bar{A} z_k + \bar{B} z_k) \\
&= z_k^T \bar{A}^T P \bar{A} z_k + 2 z_k^T \bar{B}^T P \bar{A} z_k + z_k^T \bar{B}^T P \bar{B} z_k \\
&= V_k - z_k^T Q z_k + z_k^T (2\bar{B}^T P \bar{A} + \bar{B}^T P \bar{B}) z_k
\end{aligned}
\tag{4.60}
$$

If the uncertainties are small enough such that

$$-Q + \left\| 2\bar{B}^T P \bar{A} + \bar{B}^T P \bar{B} \right\| < -\eta Q \tag{4.61}$$

for some $\eta > 0$, then

$$V_{k+1} - V_k < -\eta z_k^T Q z_k \tag{4.62}$$

It then follows that $z_k \to 0$ exponentially as $k \to \infty$. Given that x and u converge to zero at the sampling instants (see (4.50)), it follows that $u(t)$ converges to zero $\forall\, t$ as $t \to \infty$. From (4.1) it follows that $x(t)$ converges to zero $\forall\, t$ as $t \to \infty$.

4.6 Application to the Yaw Control of a Mini-helicopter

In this section we show that the proposed controller has a satisfactory behaviour when applied to control the yaw displacement of a mini-helicopter. We use a mini-helicopter with four rotors as shown in Figure 4.1 (see Chapter 3 for more details).

The platform is similar to the one we have used in the previous section (see Section 3.4), but we have made some modifications to install the new system (MaRTE OS). We will explain these modifications in detail in this section.

We aim at using visual servoing control for the mini-helicopter in future work. We know that image processing will introduce a considerable delay and one of our objectives in this chapter is to show that our prediction-based control algorithm can be used to avoid instabilities in the position and orientation control of a flying vehicle.

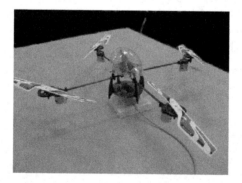

Fig. 4.1. The quad-rotor rotorcraft.

4.6.1 Characteristics of MaRTE OS

We present in this section the characteristics and implementation of the real-time control system environment that we have used. We use an embedded system based on the MaRTE OS environment.

MaRTE OS [3] is a real-time kernel for embedded applications that follows the minimal real-time POSIX.13 subset [140], providing both the C and Ada language POSIX interfaces. It allows cross-development of Ada and C real-time applications. Mixed Ada–C applications can also be developed, with a globally consistent scheduling of Ada tasks and C threads.

MaRTE OS works in a cross-development environment. The host computer is a Linux PC with the *gnat* and *gcc* compilers. The target platform is any bare machine based on any 386 PC or higher, with a floppy disk (or equivalent) for booting the application, but not requiring a hard disk.

The kernel has a low-level abstract interface for accessing the hardware. This interface encapsulates operations for interrupt management, clock and timer management and thread control.

The main applications of this kernel are industrial embedded systems developed in Ada. The hardware access facilities allow the implementation of specific device drivers in Ada style.

The final embedded system can be connected to other computers using RS232 or Ethernet drivers. Using these facilities, data can be sent to other applications to be monitored and analysed. Also, commands from other applications can be received, using the same drivers, to modify the system behaviour.

The development and execution environments are shown in Figure 4.2.

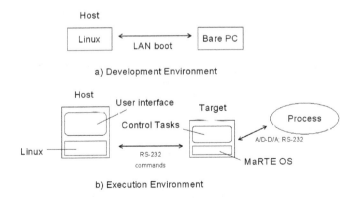

Fig. 4.2. Environment of MaRTE OS.

4.6.2 Real-time Implementation

In this subsection we present the scheme of the real-time implementation and the principal characteristics of the platform.

The mini-helicopter used is a quad-rotor rotorcraft (see Chapter 3). The radio is a *Futaba Skysport* 4. The radio and the PV are connected using data acquisition cards. In order to simplify the experiments, the control inputs can be independently commuted between the automatic and the manual control modes. The connexion in the radio in directly made to the joystick potentiometers for the gas, yaw, pitch and roll controls (see Section 3.4). The quad-rotor rotorcraft evolves freely in a 3D space without any flying stand. The position and orientation is measured using a 3D tracker system (Polhemus). The Polhemus sends this information via RS232 to the PC.

The control of the rotors is done by sending the actions to the four motors through a digital/analogue converter.

Additionally, the system will receive commands from a small keyboard and will send periodically the system status to a host to monitor the system variables and status. Figure 4.3 shows the interaction between the system and the external devices.

To design the real-time control five main tasks have been defined:

- Control_Task: this periodic task gets information about the helicopter position and calculates the actions to be sent to the motors. Considering the computation delay (between 60 and 70 ms), the period of the task is set to 80 ms. The control actions are sent to a shared protected object which stores the system information. The actions are not sent directly to the motors.

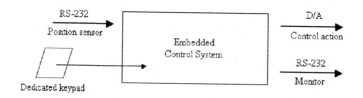

Fig. 4.3. Interaction between the system and the external devices.

- Send_Actions: this is a periodic task which is in charge of extracting the information from the control status and sending the motor actions using the digital/analogue converter. This task can introduce forced delays in the actions to be sent to the motors in order to test different control algorithms. The forced delays are introduced by getting actions calculated in previous periods when the delay is greater than the control period. If the delay is less than the period then an internal delay is executed.

- Monitor: this is a periodic task for control status monitoring. The task gets information from the shared object control status and sends it to a RS232 line to be used by the host to visualize the control variables.

- User_Commands_Task: this task reads user commands from the keyboard and executes them. User commands can change the monitoring period, change control parameters or start and stop the control.

- Control_Status: this is a shared protected object where the tasks get or put information about the process.

Several drivers have been implemented to handle the RS-232 serial line, keyboard, and the digital/analogue converters.

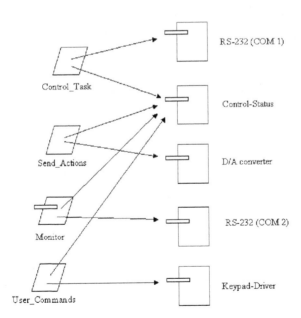

Fig. 4.4. Application architecture.

In conclusion, there are three periodic tasks: Control_Task, Send_Actions and Monitor and a sporadic task: User_Commands (Figure 4.4 and Table 4.1).

Task	Period	Priority	Offset
Control_Task	80 ms	1	0
Send_Actions	80 ms	2	10
Monitor	user_defined	3	0

Table 4.1. Periodic tasks.

The control system has been implemented in Ada and developed in a Linux-based host producing code for the target using the MaRTE OS. The image obtained is less than 300 K and runs in a bare 386 machine.

4.6.3 Experimental Results

The transfer function from the yaw-control input to the yaw-displacement has been identified by introducing a pulse input while the mini-helicopter was hovering. The obtained pulse response is shown in Figure 4.5.

We know that the mini-helicopter has an in-built gyro that introduces an angular velocity feedback. The transfer function without the gyro is basically a double integrator. However, the transfer function of the system including the angular velocity feedback, has a pole at the origin and a negative real pole.

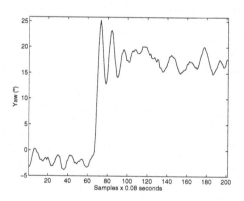

Fig. 4.5. Pulse response of the system without measurement delay.

We assumed that the system was represented by a second order system with two parameters. Trying different values for the parameters we observed that the following model has a behaviour that is close to the behaviour of the real system:

$$G(s) = \frac{200}{s(s+4)} \tag{4.63}$$

A simple controller such as

$$u_k = 0.08(y^* - y_k) \tag{4.64}$$

where y^* is a reference signal, can be used to stabilize the model (4.63). However, when there is a delay of three sampling periods (0.24 s) in the measurement of the yaw angular position, the controller becomes

$$u_k = 0.08(y^* - y_{k-3}) \tag{4.65}$$

and the closed-loop system behaviour is unstable, as can be seen in Figure 4.6.

Predictor-based system stabilization

The discrete-time state-space representation of the model in (4.63) for $T = 0.08$ s, is given by:

$$\begin{bmatrix} x_{k+1}^1 \\ x_{k+1}^2 \end{bmatrix} = \begin{bmatrix} 0.7261 & 0 \\ 0.2739 & 1 \end{bmatrix} \begin{bmatrix} x_k^1 \\ x_k^2 \end{bmatrix} + \begin{bmatrix} 0.5477 \\ 0.0923 \end{bmatrix} u_k \tag{4.66}$$

$$y_k = \begin{bmatrix} 0 & 6.25 \end{bmatrix} \begin{bmatrix} x_k^1 \\ x_k^2 \end{bmatrix} \tag{4.67}$$

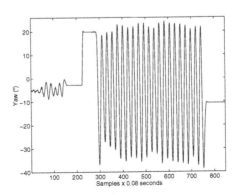

Fig. 4.6. Output of the delayed system when using the controller (4.65) without prediction.

Since the state x_k is not measurable, we use the following observer

$$\begin{bmatrix} \hat{x}^1_{k+1} \\ \hat{x}^2_{k+1} \end{bmatrix} = \begin{bmatrix} 0.7261 & -2.7940 \\ 0.2739 & -1.0261 \end{bmatrix} \begin{bmatrix} \hat{x}^1_k \\ \hat{x}^2_k \end{bmatrix} + \begin{bmatrix} 0.4470 \\ 0.3242 \end{bmatrix} y_k + \begin{bmatrix} 0.5477 \\ 0.0923 \end{bmatrix} u_{k-3} \quad (4.68)$$

The prediction of the state 3-steps ahead can be done using the following equation (see also (4.45))

$$x_p(k) = \begin{bmatrix} 0.3829 & 0 & 0.2888 & 0.3977 & 0.5477 \\ 0.6171 & 1 & 0.3512 & 0.2423 & 0.0923 \end{bmatrix} \begin{bmatrix} \hat{x}^1_k \\ \hat{x}^2_k \\ u_{k-3} \\ u_{k-2} \\ u_{k-1} \end{bmatrix} \quad (4.69)$$

The control law in (4.64) (see also (4.46)) becomes

$$u_k = 0.08(y^* - \begin{bmatrix} 0 & 6.25 \end{bmatrix} x_p(k)) \quad (4.70)$$

The yaw angular displacement of the mini-helicopter when using the above control law is shown in Figure 4.7. We have chosen y^* as a square wave function. It can be seen that the system is stabilized.

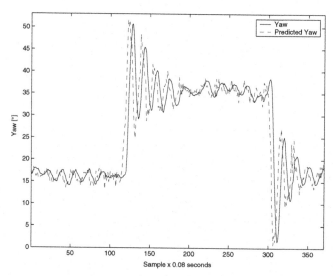

Fig. 4.7. Closed-loop behaviour using the prediction-based controller.

4.7 Conclusion

We have presented a control scheme for continuous-time systems with delay. We have proposed a discrete-time controller based on state feedback using the prediction of the state. A convergence analysis has been presented that shows that the state converges to the origin in spite of uncertainties in the knowledge of the plant parameters, the system delay and even variations of the sampling period. The proposed control scheme has been implemented to control the yaw displacement of a real four-rotor mini-helicopter. Real-time experiments have shown a satisfactory performance of the proposed control scheme.

5

Modelling and Control of Mini-helicopters

5.1 Introduction

Recently, the control of flying machines has attracted the attention of control researchers. Different approaches have been proposed to control airplanes, helicopters, rockets, satellites, etc. [103, 127, 137, 178]. Each one of these aircraft has a specific model describing its behaviour. Helicopters are among the most complex flying objects because their flight dynamics is inherently nonlinear and they have strong couplings of all the variables. The helicopter has however the ability to hover which is required in some applications. Small-scale unmanned helicopters may be expected to display considerably different dynamic response than a full-scale helicopter. The helicopter dynamic model has been modelled, in general, by Newton equations of motion. We can find in the literature complete dynamic models, as given in [78, 128, 141]; all of these contain the complete behaviour of the helicopter in different flight conditions (hover, vertical and forward flight). However, the main problem of these dynamic models is the difficulty in designing a "simple" control algorithm due to the complexity and the existence of cross-terms in the model equations. To solve this problem, some authors have proposed simple dynamic models for the controller design of these aircraft [88, 97].

The first attempt to control small helicopters was done by using linear techniques [15, 34, 88, 112, 138, 150, 151]. Nevertheless, the nonlinear nature of helicopters has to be taken into account in the controller design if one wishes to improve their performance. Backstepping techniques were used for the design of a nonlinear control law [81, 97, 51]. Recently, Isidori et al. [76] presented a nonlinear robust regulation in order to control the vertical motion of a standard helicopter. [19] uses the state-dependent Riccati equations in order to control the motion of a small helicopter and that was tested in a real application.

Generally, the study of helicopters was mainly concerned with single main rotor helicopters. Nevertheless, few studies have been conducted in order to develop controllers for tandem or coaxial rotor helicopters. NASA has started the research on flying-qualities such as directional stability, lateral oscillations, turn characteristics [86] and speed stability [170] for a tandem helicopter with nonoverlapped-rotors, with the purpose of reducing the disadvantages of this type of instability; in the same way, but for overlapped-rotors, [171] and [187] made studies on the longitudinal stability characteristics, vibrations in landing approach and yawed flight. These experiments were essential for the development of the VTOL aircraft. [155] presents a very brief control design using pole-placement theory where the feedback gains are obtained by a least squares solution, the dynamic model is based on a forward flight with linearized equations of motion. The research in [156] proposes two control laws for attitude-command and velocity-command control using digital control design and estimators. Based on the work of [156], the authors of [32] add a trajectory generator and guidance algorithms in order to obtain an autoland system. A more recent study [71] proposes two reconfigurable control laws that modify their control gains in the presence of an actuator failure in the system.

This chapter is divided as follows: Section 5.2 gives a simplified dynamic model based on Newton equations of motion for the three configurations (classic helicopter, tandem helicopter and coaxial helicopter); in Section 5.3, the helicopter dynamic model is presented in the Euler–Lagrange formulation form. In Section 5.4, both the Euler–Lagrange and Newtonian model are presented and some control algorithms are proposed. Finally, Section 5.5 presents some simulations in order to validate the model and controller proposed.

5.2 Newton–Euler Model

There exist different configurations of helicopters. The most popular are basically three: *standard configuration, tandem rotor configuration* and *coaxial rotor configuration*. We will develop the three configurations by using Newton laws. They were obtained in hover flight conditions.

5.2.1 Standard Helicopter

This model was taken basically from [36, 43]. Consider Figure 5.1. Denote $\mathcal{I} = \{E_x, E_y, E_z\}$ as a right-hand inertial frame, stationary with respect to the earth and let $\mathcal{C} = \{E_1, E_2, E_3\}$ be a right-hand body fixed frame, where $\mathcal{C}$ is fixed on the position of the centre of mass (CG) of the helicopter. In the first case, take E_z in the direction downwards into the heart, and in the second case, E_1 in the normal direction of helicopter flight, while E_3 should correspond with E_z in hover flight conditions (see Figure 5.1). Let $\mathcal{R} : \mathcal{C} \to \mathcal{I}$ be an orthogonal rotation matrix of type $\mathcal{R} \in SO(3)$:

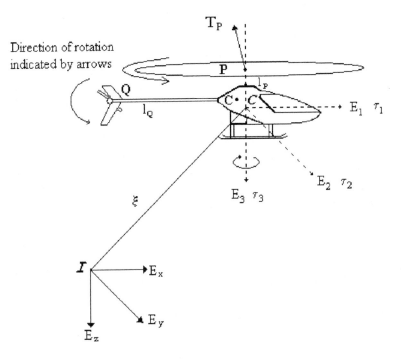

Fig. 5.1. Helicopter model representation.

$$\mathcal{R}(\eta) = \begin{pmatrix} c_\theta c_\psi & s_\phi s_\theta c_\psi - c_\phi s_\psi & c_\phi s_\theta c_\psi + s_\phi s_\psi \\ c_\theta s_\psi & s_\phi s_\theta s_\psi + c_\phi c_\psi & c_\phi s_\theta s_\psi - s_\phi c_\psi \\ -s_\theta & s_\phi c_\theta & c_\phi c_\theta \end{pmatrix} \tag{5.1}$$

denoting the helicopter orientation with respect to $\mathcal{I}$, and where $\eta = (\psi, \theta, \phi)$ describes the yaw, pitch and roll angles respectively. We have used the shorthand notation $c_\beta = \cos(\beta)$ and $s_\beta = \sin(\beta)$.

The dynamic model was obtained by taking the following assumptions:

5.1 The blades of the two rotors are considered to hinge directly from the hub, that is, there is no hinge offset associated with rotor flapping. In this way, the coning angle is assumed to be zero, As a consequence each rotor will always lie in a disk termed the rotor *disk*.

5.2 The main rotor blades are assumed to rotate in an anti-clockwise direction when viewed from above and the tail rotor blades rotate in a clockwise direction, see Figure 5.1.

5.3 It is assumed that the cyclic lateral and longitudinal tilts of the main rotor disk are measurable and controllable. That is the flapping angles are used directly as control inputs.

5.4 The only air resistance modelled are simple drag forces opposing the rotation of the two rotors.

5.5 The aerodynamic forces generated by the relative wind are not considered.

5.6 The effects of operating the helicopter close to the ground are neglected.

5.7 The effects of the aerodynamic forces generated by the stabilizers are not taken into account.

In order to obtain the dynamic equations, we have separated the aerodynamic forces into two groups. The first group is composed of translational forces and the second is related to the rotational forces of motion. More details on helicopter dynamics can be found in [78, 128, 141].

Translational forces

We found three types of forces applied to the fuselage. Two of them are generated by the two rotors. The last is due to the gravity effect. Denote by M and T the terms related with the main and tail rotor respectively. The vector thrust of the main and tail rotors are described by

$$T_M = T_M^1 E_1 + T_M^2 E_2 + T_M^3 E_3 \tag{5.2}$$
$$T_T = T_T^1 E_1 + T_T^2 E_2 + T_T^3 E_3 \tag{5.3}$$

However, it is well known that the tail rotor has no swashplate. Then, the thrust vector of this rotor always has the same direction, i.e. in the direction of the E_2 axis, so equation (5.3) can be rewritten as

$$T_T = T_T^2 E_2 \tag{5.4}$$

The components of the thrust vector for the main rotor can be defined as a function of an angle β called the *flapping angle* which denotes the tilt of the main rotor disk with respect to its initial rotation plane (Figure 5.2). This angle is formed by the tilts of angle a (longitudinal flapping) and angle b (lateral flapping) that we have assumed to be measurable and controllable variables (Assumption 5.3). By simple geometric calculus, we have

$$\tan^2 \beta = \frac{(T_M^1)^2 + (T_M^2)^2}{(T_M^3)^2} \tag{5.5}$$

$$\tan^2 \beta = \tan^2 a + \tan^2 b \tag{5.6}$$

$$\frac{1 - \cos^2 \beta}{\cos^2 \beta} = \frac{\sin^2 a}{\cos^2 a} + \frac{\sin^2 b}{\cos^2 b} \tag{5.7}$$

Simplifying the last equation, we obtain the expression

$$\cos \beta = \frac{\cos a \cdot \cos b}{\sqrt{1 - \sin^2 a \cdot \sin^2 b}} \tag{5.8}$$

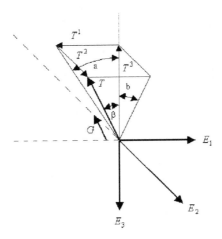

Fig. 5.2. Thrust vector of the main rotor.

The term T_M^3 is directly obtained by the projection of the thrust vector on the axis E_3. We add a negative sign to this equation because we assume that the amplitude of T_M is always positive, i.e. in the opposite direction to the axis E_3:

$$
\begin{aligned}
T_M^3 &= -\cos\beta|T_M| \\
&= \frac{-\cos a \cdot \cos b}{\sqrt{1 - \sin^2 a \cdot \sin^2 b}} \cdot |T_M|
\end{aligned}
\tag{5.9}
$$

Then, the terms T_M^1 and T_M^2 are obtained in a similar way, so

$$
\begin{aligned}
T_M^1 &= \tan a \cdot T_M^3 \\
&= \frac{-\sin a \cdot \cos b}{\sqrt{1 - \sin^2 a \cdot \sin^2 b}} \cdot |T_M|
\end{aligned}
\tag{5.10}
$$

and

$$
\begin{aligned}
T_M^2 &= -\tan b \cdot T_M^3 \\
&= \frac{\cos a \cdot \sin b}{\sqrt{1 - \sin^2 a \cdot \sin^2 b}} \cdot |T_M|
\end{aligned}
\tag{5.11}
$$

The thrust vector T_M can be represented as

$$
T_M = G(a,b) \cdot |T_M|
\tag{5.12}
$$

where

$$
G(a,b) = \frac{1}{\sqrt{1 - \sin^2 a \cdot \sin^2 b}} \begin{pmatrix} -\sin a \cdot \cos b \\ \sin b \cdot \cos a \\ -\cos a \cdot \cos b \end{pmatrix}
\tag{5.13}
$$

Remark 5.1. Whenever the values of the angles a and b are very small, the equation of $G(a, b)$ can be rewritten as

$$G(a, b) \approx \begin{pmatrix} -a \\ b \\ -1 \end{pmatrix} \tag{5.14}$$

The gravitational force applied to the helicopter is defined as

$$f_g = mgE_z \tag{5.15}$$

where m represents the total mass of the helicopter and g is the gravity constant. This equation is given on the inertial fixed frame $\mathcal{I}$. For the body frame $\mathcal{C}$, we have

$$F_g = mgR^T E_z \tag{5.16}$$

Let F be the external total force applied to the helicopter and expressed on the body fixed frame. This force is given by

$$\begin{aligned} F &= T_M + T_T + F_g \\ &= G(a, b) \cdot |T_M| + T_T^2 E_2 + mgR^T E_z \end{aligned} \tag{5.17}$$

The representation of this force on the inertial fixed frame has the form

$$\begin{aligned} f &= \mathcal{R}F \\ &= \mathcal{R}G(a, b) \cdot |T_M| + T_T^2 \mathcal{R}E_2 + mgE_z \end{aligned} \tag{5.18}$$

Torques

The two thrust vectors T_M and T_T generate two torques due to separation between the centre of the mass CG and the rotor hubs. The gravitational force does not generate a torque since the helicopter is free to rotate around its centre of mass. Before beginning, it is necessary to define the measured distances from the centre of mass of the helicopter to the hubs of the two rotors (denoted l_M for the main rotor and l_T for the tail rotor). If we express these vectors in terms of the body fixed frame, we have

$$l_M = l_M^1 E_1 + l_M^2 E_2 + l_M^3 E_3 \tag{5.19}$$
$$l_T = l_T^1 E_1 + l_T^2 E_2 + l_T^3 E_3 \tag{5.20}$$

The torques applied to the airframe by the thrust vectors are defined by

$$\begin{aligned} \tau_M &= [l_M \times G(a, b)]|T_M| \\ &= \frac{|T_M|}{\sqrt{1 - \sin^2 a \cdot \sin^2 b}} \begin{pmatrix} -\cos a \cdot \cos b \cdot l_M^2 - \sin b \cdot \cos a \cdot l_M^3 \\ \cos a \cdot \cos b \cdot l_M^1 - \sin a \cdot \cos b \cdot l_M^3 \\ \sin a \cdot \cos b \cdot l_M^2 + \sin b \cdot \cos a \cdot l_M^1 \end{pmatrix} \end{aligned} \tag{5.21}$$

$$\begin{aligned} \tau_T &= [l_T \times E_2]T_T^2 \\ &= T_T^2 \begin{pmatrix} -l_T^3 \\ 0 \\ l_T^1 \end{pmatrix} \end{aligned} \tag{5.22}$$

Additionally, the aerodynamic drags on the rotors generate some pure torques acting through the rotor hubs. Evoking Assumption 5.2, these anti-torques are defined by

$$Q_M = |Q_M|E_3 \tag{5.23}$$
$$Q_T = -|Q_T|E_2 \tag{5.24}$$

Finally, the total torque applied to the fuselage C, represented in the body fixed frame, is given by

$$\tau = [l_M \times G(a,b)]|T_M| + [l_T \times E_2]T_T^2 + |Q_M|E_3 - |Q_T|E_2 \tag{5.25}$$

Complete dynamic model

For the translational motion of the helicopter, let $\dot{\xi} = v$ denote the velocity of its centre of mass expressed in the inertial frame $\mathcal{I}$. Newton's equations show that the rotational component of motion in a non-inertial frame is given by $\mathbf{I}\dot{\Omega} = -\Omega \times \mathbf{I}\Omega + \tau$, where Ω is the angular velocity expressed in the non-inertial frame; $\mathbf{I}$ denote the inertia of the helicopter around its centre of mass with respect to the body fixed frame and τ is the applied external torque in the body fixed frame. Finally, the full dynamic model, represented in the inertial fixed frame, is given by

$$\dot{\xi} = v \tag{5.26}$$
$$m\dot{v} = \mathcal{R}G(a,b) \cdot |T_M| + T_T^2 \mathcal{R}E_2 + mgE_z \tag{5.27}$$
$$\dot{\mathcal{R}} = \mathcal{R}\hat{\Omega} \tag{5.28}$$
$$\mathbf{I}\dot{\Omega} = -\Omega \times \mathbf{I}\Omega + [l_M \times G(a,b)]|T_M| + [l_T \times E_2]T_T^2$$
$$+ |Q_M|E_3 - |Q_T|E_2 \tag{5.29}$$

where $\Omega \in \mathbb{R}^3$ and

$$\hat{\Omega} = \begin{pmatrix} 0 & -\Omega^3 & \Omega^2 \\ \Omega^3 & 0 & -\Omega^1 \\ -\Omega^2 & \Omega^1 & 0 \end{pmatrix} \tag{5.30}$$

5.2.2 Tandem Helicopter

A tandem rotor helicopter (Figure 5.3) uses two contrarotating rotors of equal size and loading, so there is no net yaw moment on the helicopter because the torques of the rotors are equal and opposing. Typically, the two rotors are overlapped by around 20% to 50% of the radius (r) of the rotor disk, so the shaft separation is thus around 1.8r to 1.5r. To minimize the aerodynamic interference created by the operation of the rear rotor in the wake of the front, the rear rotor is elevated on a pylon (0.3r to 0.5r above the front rotor).

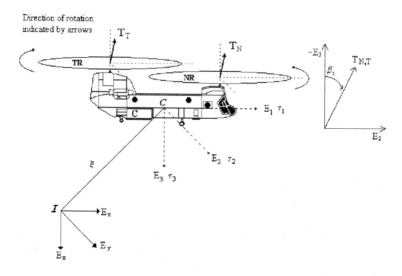

Fig. 5.3. A tandem helicopter configuration.

In a tandem rotor helicopter, pitch moment is achieved by differential change of the main rotors thrust magnitude (by collective pitch), roll moment is controlled by lateral thrust tilt using cyclic pitch (Figure 5.4), yaw moment is obtained by differential lateral tilt of the thrust on the two main rotors with cyclic pitch (Figure 5.5), and the vertical force is achieved by the change of the main rotor collective pitch.

For simplicity we will present here the dynamic model of a tandem main rotor helicopter in hovering. We propose a dynamic tandem helicopter model based on Newton's equations of motion [59] with the assumptions of the standard helicopter with the following changes:

1T The nose rotor blades are assumed to rotate in an anti-clockwise direction when viewed from above and the tail rotor blades rotate in a clockwise direction, see Figure 5.3.

2T The operation of two or more rotors in close proximity will modify the flow field at each, and hence the performance of the rotor system will not be the same as for the isolated rotors. We will not consider this phenomenon to simplify the dynamical model.

In order to obtain the final dynamic equations, we have separated the aerodynamic forces into two groups. The first group is composed of translational forces and the second is related to the rotational forces of motion. More details on tandem rotor helicopter dynamics can be found in [78].

Both Rotor Systems Tilted
to the Right

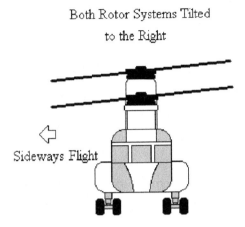

Sideways Flight

Fig. 5.4. A tandem helicopter in sideways flight.

Pivot Around the Center of the Helicopter

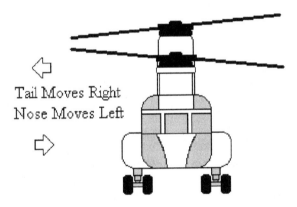

Tail Moves Right
Nose Moves Left

Fig. 5.5. Control direction for a tandem helicopter.

Translational forces

Denote by T_N and T_T the thrust generated by the nose (N) and tail (T) rotors respectively (Figure 5.3). These forces have no E_1 component, so the thrust vectors are defined by

$$T_N = T_N^2 E_2 - T_N^3 E_3, \qquad T_T = T_T^2 E_2 - T_T^3 E_3 \qquad (5.31)$$

By simple geometric analysis we obtain expressions in terms of β, where β is the angle between the axis E_3 and the actual thrust vector:

$$T_N = |T_N| \sin \beta_N E_2 - |T_N| \cos \beta_N E_3 \qquad (5.32)$$

$$T_T = |T_T| \sin \beta_T E_2 - |T_T| \cos \beta_T E_3 \qquad (5.33)$$

The thrust vector can be represented by the expression

$$T_i = |T_i| \begin{bmatrix} 0 \\ \sin \beta_i \\ -\cos \beta_i \end{bmatrix} = |T_i| \begin{bmatrix} 0 \\ \beta_i \\ -1 \end{bmatrix} \qquad (5.34)$$

where $i = N$ or T, and considering sufficiently small values of β_i. Another force applied to the tandem rotor helicopter is the gravitational force given by $f_g = mgE_z$, where m is the complete mass of the helicopter and g is the gravitational constant. The above expression is defined in the inertial frame $\mathcal{I}$. In terms of the body fixed frame, it is necessary to multiply f_g by the inverse of the rotation matrix $\mathcal{R}$ that represents the orientation of the body-fixed frame $\mathcal{C}$ with respect to $\mathcal{I}$.

Denote by f the total translational force applied to the helicopter expressed in the inertial frame $\mathcal{I}$

$$f = (|T_N|\beta_N + |T_T|\beta_T)\mathcal{R}E_2 - (|T_N| + |T_T|)\mathcal{R}E_3 + mgE_z \qquad (5.35)$$

Torques and anti-torques

The torques generated by the thrust vectors T_N and T_T are due to separation between the centre of mass CG and the rotor hubs (called τ_N and τ_T respectively). The gravitational force does not generate a torque since the helicopter is free to rotate around its centre of mass. Before beginning, it is necessary to define the measured distances from the centre of mass of the tandem rotor helicopter to the hubs of the two rotors (denoted l_N for the nose rotor and l_T for the tail rotor). If we express these vectors in terms of the body fixed frame, we have

$$l_N = l_N^1 E_1 - l_N^3 E_3 \qquad (5.36)$$

$$l_T = -l_T^1 E_1 - l_T^3 E_3 \qquad (5.37)$$

The torques applied to the airframe by the thrust vectors are defined by

$$\tau_N = l_N \times T_N$$
$$\tau_T = l_T \times T_T \qquad (5.38)$$

The total torque generated by the nose and tail rotors is given by

$$\tau_{NT} = \tau_N + \tau_T$$

$$= \begin{bmatrix} l_N^3 |T_N| \beta_N + l_T^3 |T_T| \beta_T \\ l_N^1 |T_N| - l_T^1 |T_T| \\ l_N^1 |T_N| \beta_N - l_T^1 |T_T| \beta_T \end{bmatrix} \tag{5.39}$$

Additionally, the aerodynamic drags on the rotors generate some pure torques acting through the rotor hubs. Evoking Assumption 2T, the anti-torques are defined by

$$Q_N = |Q_N| E_3, \quad Q_T = -|Q_T| E_3 \tag{5.40}$$

Finally, the total torque applied to the tandem rotor helicopter (expressed in the body fixed frame) is given by

$$\tau = \tau_{NT} + |Q_N| E_3 - |Q_T| E_3 \tag{5.41}$$

Complete model

Substituting the total force (5.35) and total torque in a Newton basic model, we have

$$\dot{\xi} = v \tag{5.42}$$

$$m\dot{v} = (|T_N| \beta_N + |T_T| \beta_T) \mathcal{R} E_2 - (|T_N| + |T_T|) \mathcal{R} E_3 + mg E_z \tag{5.43}$$

$$\dot{\mathcal{R}} = \mathcal{R} \hat{\Omega} \tag{5.44}$$

$$\mathbf{I}\dot{\Omega} = -\Omega \times \mathbf{I}\Omega + |Q_N| E_3 - |Q_T| E_3 + \tau_{NT} \tag{5.45}$$

5.2.3 Coaxial Helicopter

A coaxial helicopter is again a twin main rotor helicopter configuration that uses two contrarotating rotors of equal size and loading and with concentric shafts (see Figure 5.6). Some vertical separation of the rotor disks is required to accommodate lateral flapping. Pitch and roll control is achieved by the cyclic pitch of the swash plate. The height control is achieved by the collective pitch. The yaw control mechanism is more subtle. When a rotor turns, it has to overcome air resistance, so a reactive force acts on the rotor in the direction opposite to the rotation of the rotor. As long as all rotors produce the same torque, they produce the same reactive torque. This torque is mostly a function of rotation speed and rotor blade pitch.

Since the sum of the two air resistances is zero, there is no yaw motion. If one of the rotors changes its collective pitch, the induced torque will cause the helicopter to rotate in the direction of the induced torque. It is important to note that this operation has no effect on translation in the x or y direction in a coaxial helicopter configuration. For the sake of simplicity we present here the dynamic model of a coaxial helicopter in hovering. We use the same assumptions taken for the standard helicopter, adding/changing only the following assumptions:

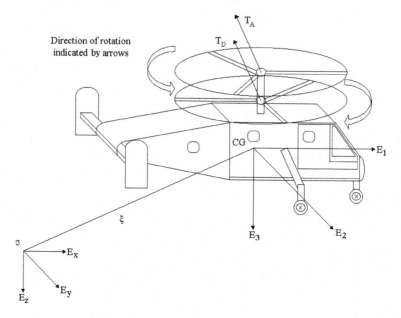

Fig. 5.6. Coaxial helicopter configuration.

1C The above rotor blades are assumed to rotate in an anti-clockwise direction when viewed from above and the down rotor blades rotate in a clockwise direction, see Figure 5.6.

2C The operation of two or more rotors in close proximity will modify the flow field at each, and hence the performance of the rotor system will not be the same as for the isolated rotors. We will not consider this phenomenon to simplify the dynamic model.

We have separated the aerodynamic forces into two parts. The first part is composed of the total translational forces applied to the coaxial helicopter and the second part is related to the sum of the rotational torques. More details on coaxial helicopter dynamics can be found in [78].

Translational forces

Denote by T_A and T_D the thrusts generated by the above (A) and down (D) rotors respectively. Then, the thrust vectors are defined by

$$T_A = T_A^1 E_1 + T_A^2 E_2 - T_A^3 E_3 \tag{5.46}$$
$$T_D = T_D^1 E_1 + T_D^2 E_2 - T_D^3 E_3 \tag{5.47}$$

It is desirable to express the thrust components in terms of the cyclic tilt angles a and b which form the system inputs. Define β as the measure of the tilt of the rotor disk (see Figure 5.2). Thus, the thrust components can be directly computed by the projection of T_A or T_D onto the axis of the body fixed frame by

$$T_A = G(a, b) \cdot |T_A| \tag{5.48}$$
$$T_D = G(a, b) \cdot |T_D| \tag{5.49}$$

where $G(a, b)$ are defined in equation (5.13). Another force applied to the coaxial helicopter is the gravitational force given by (5.15). Denote by f the total translational force applied to the helicopter and expressed in the inertial-fixed frame:

$$f = \mathcal{R}G(a, b)(|T_A| + |T_D|) + mgE_3 \tag{5.50}$$

Torques and anti-torques

The torques generated by the thrust vectors T_A and T_D, represented by τ_A and τ_D respectively, are due to separation between the centre of mass and the rotor hubs. The gravitational force does not generate a torque since the helicopter is free to rotate around its centre of mass.

Define the measured distances from the centre of mass to the hubs of the two rotors as l_A for the above rotor and l_D for the down rotor. The torques applied to the airframe by the thrust vectors are defined by

$$\tau_A = l_A \times T_A \tag{5.51}$$
$$\tau_D = l_D \times T_D \tag{5.52}$$

The total torque generated by the above and down rotors is given by the sum of equations (5.51) and (5.52); it is given by

$$\tau_{AD}^1 = -(l_A^3 |T_A| + l_D^3 |T_D|)G^2(a, b) + (l_A^2 |T_A| + l_A^2 |T_D|)G^3(a, b) \tag{5.53}$$
$$\tau_{AD}^2 = (l_A^3 |T_A| + l_D^3 |T_D|)G^1(a, b) - (l_A^1 |T_A| + l_D^1 |T_D|)G^3(a, b) \tag{5.54}$$
$$\tau_{AD}^3 = -(l_A^2 |T_A| + l_D^2 |T_D|)G^1(a, b) + (l_A^1 |T_A| + l_D^1 |T_D|)G^2(a, b) \tag{5.55}$$

The aerodynamic drags on the rotors generate some pure torques acting through the rotor hubs. Then, the anti-torques are defined by

$$Q_A = |Q_A|E_3 \tag{5.56}$$
$$Q_D = -|Q_D|E_3 \tag{5.57}$$

Finally, the total torque applied to the coaxial helicopter, expressed in the body fixed frame, is given by

$$\tau = \tau_{AD} + |Q_A|E_3 - |Q_D|E_3 \tag{5.58}$$

Complete model

Substituting the total force (5.50) and total torque (5.58) in a Newton basic model, we have

$$\dot{\xi} = v \tag{5.59}$$
$$m\dot{v} = \mathcal{R}G(a,b)(|T_A| + |T_D|) + mgE_3 \tag{5.60}$$
$$\dot{\mathcal{R}} = \mathcal{R}\hat{\Omega} \tag{5.61}$$
$$\mathbf{I}\dot{\Omega} = -\Omega \times \mathbf{I}\Omega + |Q_A|E_3 - |Q_D|E_3 + \tau_{AD} \tag{5.62}$$

5.2.4 Adapted Dynamic Model for Control Design

Comparing the three dynamic models (standard, tandem and coaxial), we can define a general dynamic model that separates the main force for sustaining the helicopter from the small-body forces. Taking advantage of this change, we can also adapt the model in order to have a simple dynamic model for control design use.

Consider the case of the standard helicopter where the translational force applied to the fuselage, without the gravity term and in hovering conditions, is given by

$$f = \mathcal{R} \cdot |T_M| \begin{pmatrix} -aE_1 \\ bE_2 \\ E_3 \end{pmatrix} + T_T^2 \mathcal{R}E_2 \tag{5.63}$$

Recall the equation of the torques generated by the two rotors

$$\tau = [l_M \times G(a,b)]|T_M| + [l_T \times E_2]T_T^2$$
$$= |T_M| \begin{pmatrix} 0 & -l_M^3 & l_M^2 \\ l_M^3 & 0 & -l_M^1 \\ -l_M^2 & l_M^1 & 0 \end{pmatrix} \begin{pmatrix} G^1(a,b) \\ G^2(a,b) \\ G^3(a,b) \end{pmatrix} + \begin{pmatrix} -l_T^3 \\ 0 \\ l_T^1 \end{pmatrix} T_T^2 \tag{5.64}$$

where $G^1(a,b), G^2(a,b)$ and $G^3(a,b)$ are the terms of the vector $G(a,b) \in \mathbb{R}^3$. In (5.64), taking the first two columns of the first term along with the second term into a single matrix yields

$$\tau = \begin{pmatrix} 0 & -l_M^3 & l_T^3 \\ l_M^3 & 0 & 0 \\ -l_M^2 & l_M^1 & l_T^1 \end{pmatrix} \begin{pmatrix} |T_M|G^1(a,b) \\ |T_M|G^2(a,b) \\ T_T^2 \end{pmatrix}$$
$$+ \begin{pmatrix} l_M^2 \\ -l_M^1 \\ 0 \end{pmatrix} |T_M|G^3(a,b) \tag{5.65}$$

In order to simplify the representation of the last equation, define

$$K = \begin{pmatrix} 0 & -l^3_M & l^3_T \\ l^3_M & 0 & 0 \\ -l^2_M & l^1_M & l^1_T \end{pmatrix}, \quad k_0 = \begin{pmatrix} l^2_M \\ -l^1_M \\ 0 \end{pmatrix} \qquad (5.66)$$

It easy to see that the term involving K contributes significantly to the control of the rotation dynamics, while the term involving k_0 is a coupling between the translational force control and the rotation dynamics. Then, we introduce here a main vector of control inputs for the rotation dynamics

$$\gamma = \begin{pmatrix} \gamma^1 \\ \gamma^2 \\ \gamma^3 \end{pmatrix} = K \begin{pmatrix} |T_M|G^1(a,b) \\ |T_M|G^2(a,b) \\ T^2_T \end{pmatrix} \qquad (5.67)$$

It is important to note that the parameters $l^3_M \gg l^1_M, l^2_M$ and $l^1_T \gg l^3_T$, thus we can verify that the matrix K is clearly full rank. Moreover, in normal flight conditions, the value of $|T_M| \gg 0$, then small changes in the cycle tilt angles a and b along with the tail thrust T^2_T will allow arbitrary (though bounded) control action γ.

Consider the translational force given in (5.63). This force can be rewritten in the body-fixed frame as

$$F = |T_M|G^3(a,b)E_3 + [E_1 E_2 E_2] \begin{pmatrix} |T_M|G^1(a,b) \\ |T_M|G^2(a,b) \\ T^2_T \end{pmatrix} \qquad (5.68)$$

Define a nominal control ($u > 0$) associated with the translation dynamics in

$$u = -|T_M|G^3(a,b) = -T^3_M \qquad (5.69)$$

where the negative sign is due to the direction of $|T_M|$, that is opposed to E_3.

Let

$$L = \begin{pmatrix} 1 & 0 & 0 \\ 0 & 1 & 1 \\ 0 & 0 & 0 \end{pmatrix} \qquad (5.70)$$

and note that $[E_1 E_2 E_2] = \mathcal{R}L$. Moreover, we can verify that $E_3\mathcal{R}L = 0$. Using (5.67) and (5.69), we have

$$F = -uE_3 + \mathcal{R}LK^{-1}\gamma$$
$$= -uE_3 + \mathcal{R}\sigma\gamma \qquad (5.71)$$

where $\sigma = LK^{-1}$. Finally, the translational force can be written with two terms: the first (u) is related to the main control input for the translation dynamics; the second term (σ) is orthogonal to E_3 and is expected to be of a much smaller magnitude. It corresponds to the small body forces exerted on the airframe when torque control is applied. Recalling equation (5.65) and using (5.66), it follows that

$$\Gamma = \gamma + k_0 u \tag{5.72}$$

This equation shows the full torque control available via the control input γ. It also describes the existing coupling between translational and rotation inputs expressed as $k_0 u$. If we suppose that the axis E_z is placed lower down the rotor hub, then $k_0 = 0$.

The equivalent dynamic model, then is given by

$$\dot{\xi} = v \tag{5.73}$$
$$m\dot{v} = -u\mathcal{R}E_3 + mgE_z + \mathcal{R}\sigma\gamma \tag{5.74}$$
$$\dot{\mathcal{R}} = \mathcal{R}\hat{\Omega} \tag{5.75}$$
$$\mathbf{I}\dot{\Omega} = -\Omega \times \mathbf{I}\Omega + |Q_M|E_3 - |Q_T|E_2 + \Gamma \tag{5.76}$$

The block diagram of the above model can be represented as shown in Figure 5.7.

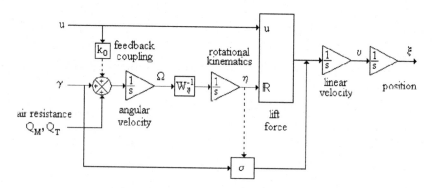

Fig. 5.7. Block diagram of the dynamic model of a standard helicopter.

General dynamic model

We propose the following dynamic model for the three helicopters:

$$\dot{\xi} = v \tag{5.77}$$
$$m\dot{v} = -u\mathcal{R}E_3 + mgE_z + \mathcal{R}\sigma\gamma \tag{5.78}$$
$$\dot{\mathcal{R}} = \mathcal{R}\hat{\Omega} \tag{5.79}$$
$$\mathbf{I}\dot{\Omega} = -\Omega \times \mathbf{I}\Omega + \Gamma_Q + \Gamma \tag{5.80}$$

where Γ_Q is the total anti-torques due to the air resistance. For a standard helicopter, u is given by equation (5.69), Γ is described in equation (5.72) and $\Gamma_Q = |Q_M|E_3 - |Q_T|E_2$.

For a tandem helicopter, we have

$$u = -(|T_N| + |T_T|)E_3 \tag{5.81}$$

$$\sigma\gamma = (|T_N|\beta_N + |T_T|\beta_T)E_2 \tag{5.82}$$

$$\Gamma = \tau_{NT} \tag{5.83}$$

$$\Gamma_Q = |Q_N|E_3 - |Q_T|E_3 \tag{5.84}$$

where

$$\beta_N = \frac{l_T^1 \tau_{NT}^1 + l_T^3 \tau_{NT}^3}{(l_T^1 l_N^3 + l_T^3 l_N^1)|T_N|} \tag{5.85}$$

$$\beta_T = \frac{\tau_{NT}^1 - l_N^3 \tau_{NT}^3}{l_T^3 |T_T|} \tag{5.86}$$

Finally, for a coaxial helicopter, it yields

$$u = -(|T_A| + |T_D|)G^3(a,b) \tag{5.87}$$

$$\sigma\gamma = (|T_A| + |T_D|)\begin{pmatrix} G^1(a,b) \\ G^2(a,b) \\ 0 \end{pmatrix} \tag{5.88}$$

$$\Gamma = \tau_{AD} \tag{5.89}$$

$$\Gamma_Q = |Q_A|E_3 - |Q_D|E_3 \tag{5.90}$$

where

$$a = -\frac{\tau_{AD}^2 - l_A^1|T_A| - l_D^1|T_D|}{l_A^3|T_A| + l_D^3|T_D|} \tag{5.91}$$

$$b = -\frac{\tau_{AD}^1 + l_A^2|T_A| + l_D^2|T_D|}{l_A^3|T_A| + l_D^3|T_D|} \tag{5.92}$$

5.3 Euler–Lagrange Model

Another way to represent a dynamic model is by using the equations of motion of Euler–Lagrange. Here, the model of [9, 43, 94] works. Define the generalized coordinates of the helicopter as

$$q = (\xi, \eta)^T = (x, y, z, \psi, \theta, \phi)^T \in \mathbb{R}^6 \tag{5.93}$$

where ξ and η represent the position and orientation of the helicopter with respect to the inertial-fixed frame respectively (see Figure 5.1).

The translational and rotational kinetic energy of the helicopter are

$$T_{trans} = \frac{m}{2} \left\langle \dot{\xi}, \dot{\xi} \right\rangle$$
$$= \frac{m}{2} (\dot{x}^2 + \dot{y}^2 + \dot{z}^2) \tag{5.94}$$

$$T_{rot} = \frac{1}{2} \Omega^T \mathbf{I} \Omega \tag{5.95}$$

The angular velocity in the body-fixed frame C is related to the generalized velocities $(\dot{\psi}, \dot{\theta}, \dot{\phi})$ [59]:

$$\Omega_a = \begin{pmatrix} \dot{\phi} - \dot{\psi} s_\theta \\ \dot{\theta} c_\phi + \dot{\psi} c_\theta s_\phi \\ \dot{\psi} c_\theta c_\phi - \dot{\theta} s_\phi \end{pmatrix} \tag{5.96}$$

which can also be written as

$$\Omega_a = W_\eta \dot{\eta} \tag{5.97}$$

where

$$W_\eta = \begin{pmatrix} -s_\theta & 0 & 1 \\ c_\theta s_\phi & c_\phi & 0 \\ c_\theta c_\phi & -s_\phi & 0 \end{pmatrix} \tag{5.98}$$

Therefore

$$\dot{\eta} = \begin{pmatrix} \dot{\psi} \\ \dot{\theta} \\ \dot{\phi} \end{pmatrix} = W_\eta^{-1} \Omega_a \tag{5.99}$$

The total kinetic energy of the system is given by

$$T = T_{trans} + T_{rot} = \frac{1}{2} M \left\langle \dot{\xi}, \dot{\xi} \right\rangle + \frac{1}{2} \left\langle \Omega_a, \mathbf{I}_C \Omega_a \right\rangle \tag{5.100}$$

where M is the total mass of the helicopter, g is the gravitation constant and $\mathbf{I}_C$ denotes the inertia of the airframe C:

$$\mathbf{I}_C = \begin{bmatrix} I_1 & 0 & 0 \\ 0 & I_2 & 0 \\ 0 & 0 & I_3 \end{bmatrix} \tag{5.101}$$

The potential energy is

$$U = -Mgz \tag{5.102}$$

and thus the Lagrangian function is defined as

$$L = T + U \tag{5.103}$$

which satisfies the Euler–Lagrange equation

$$\frac{d}{dt} \left(\frac{\partial L}{\partial \dot{q}} \right) - \frac{\partial L}{\partial q} = \mathbf{F_L} \tag{5.104}$$

where $\mathbf{F_L} = (f, \tau)$ represents the forces and torques applied to the fuselage. Since the Lagrangian contains no cross-terms in the kinetic energy combining $\dot{\xi}$ with $\dot{\eta}$ the Euler–Lagrange equation (5.104) can be partitioned into dynamics for ξ coordinates and η coordinates. Then

$$L_{trans} = \frac{M}{2}(\dot{x}^2 + \dot{y}^2 + \dot{z}^2) - Mgz \qquad (5.105)$$

The Euler–Lagrange equation for the translation motion is

$$\frac{d}{dt}\left[\frac{\partial L_{trans}}{\partial \dot{q}}\right] - \frac{\partial L_{trans}}{\partial q} = F_\xi \qquad (5.106)$$

where

$$F_\xi = \begin{bmatrix} M\ddot{x} \\ M\ddot{y} \\ M\ddot{z} - Mg \end{bmatrix} \qquad (5.107)$$

As for the η coordinates, we can rewrite

$$\frac{d}{dt}\left[(\Omega_a)^T \mathbb{I}_C \frac{\partial \Omega_a}{\partial \dot{q}}\right] - (\Omega_a)^T \mathbb{I}_C \frac{\partial \Omega_a}{\partial q} = F(\tau) \qquad (5.108)$$

where

$$\frac{\partial \Omega_a}{\partial \dot{q}} = \begin{bmatrix} -s_\theta & 0 & 1 \\ c_\theta s_\phi & c_\phi & 0 \\ c_\theta c_\phi & -s_\phi & 0 \end{bmatrix} \qquad (5.109)$$

$$(\Omega_a)^T \mathbb{I}_C \frac{\partial \Omega_a}{\partial \dot{q}} = \begin{bmatrix} b_1 & b_2 & b_3 \end{bmatrix} \qquad (5.110)$$

$$b_1 = -I_1(\dot{\phi}s_\theta - \dot{\psi}s_\theta^2) + I_2(\dot{\theta}c_\theta s_\phi c_\phi + \dot{\psi}c_\theta^2 s_\phi^2)$$
$$+ I_3(\dot{\psi}c_\theta^2 c_\phi^2 - \dot{\theta}c_\theta s_\phi c_\phi) \qquad (5.111)$$

$$b_2 = I_2(\dot{\theta}c_\phi^2 + \dot{\psi}c_\theta s_\phi c_\phi) - I_3(\dot{\psi}c_\theta s_\phi c_\phi - \dot{\theta}s_\phi^2) \qquad (5.112)$$

$$b_3 = I_1(\dot{\phi} - \dot{\psi}s_\theta) \qquad (5.113)$$

and

$$\frac{d}{dt}\left[(\Omega_a)^T \mathbb{I}_C \frac{\partial \Omega_a}{\partial \dot{q}}\right] = \begin{bmatrix} d_1 & d_2 & d_3 \end{bmatrix} \qquad (5.114)$$

where

$$d_1 = -I_1(\ddot{\phi}s_\theta + \dot{\phi}\dot{\theta}c_\theta - \ddot{\psi}s_\theta^2 - 2\dot{\psi}\dot{\theta}s_\theta c_\theta)$$
$$+ I_2(\ddot{\theta}c_\theta s_\phi c_\phi - \dot{\theta}^2 s_\theta s_\phi c_\phi - \dot{\theta}\dot{\phi}c_\theta s_\phi^2 + \dot{\theta}\dot{\phi}c_\theta c_\phi^2$$
$$+ \ddot{\psi}c_\theta^2 s_\phi^2 - 2\dot{\psi}\dot{\theta}s_\theta c_\theta s_\phi^2 + 2\dot{\psi}\dot{\phi}c_\theta^2 s_\phi c_\phi)$$
$$+ I_3(\ddot{\psi}c_\theta^2 c_\phi^2 - 2\dot{\psi}\dot{\theta}s_\theta c_\theta c_\phi^2 - 2\dot{\psi}\dot{\phi}c_\theta^2 s_\phi c_\phi$$
$$- \ddot{\theta}c_\theta s_\phi c_\phi + \dot{\theta}^2 s_\theta s_\phi c_\phi + \dot{\theta}\dot{\phi}c_\theta s_\phi^2 - \dot{\theta}\dot{\phi}c_\theta c_\phi^2) \qquad (5.115)$$

$$d_2 = I_2(\ddot{\theta}c_\phi^2 - 2\dot{\theta}\dot{\phi}s_\phi c_\phi + \ddot{\psi}c_\theta s_\phi c_\phi - \dot{\psi}\dot{\theta}s_\theta s_\phi c_\phi + \dot{\psi}\dot{\phi}c_\theta c_\phi^2$$
$$- \dot{\psi}\dot{\phi}c_\theta s_\phi^2) - I_3(\ddot{\psi}c_\theta s_\phi c_\phi - \dot{\psi}\dot{\theta}s_\theta s_\phi c_\phi - \dot{\psi}\dot{\phi}c_\theta s_\phi^2$$
$$+ \dot{\psi}\dot{\phi}c_\theta c_\phi^2 - \ddot{\theta}s_\phi^2 - 2\dot{\theta}\dot{\phi}s_\phi c_\phi) \tag{5.116}$$

$$d_3 = I_1(\ddot{\phi} - \ddot{\psi}s_\theta - \dot{\psi}\dot{\theta}c_\theta) \tag{5.117}$$

Also

$$\frac{\partial \Omega_a}{\partial q} = \begin{bmatrix} 0 & -\dot{\psi}c_\theta & 0 \\ 0 & -\dot{\psi}s_\theta s_\phi & -\dot{\theta}s_\phi + \dot{\psi}c_\theta c_\phi \\ 0 & -\dot{\psi}s_\theta c_\phi & -\dot{\psi}c_\theta s_\phi - \dot{\theta}c_\phi \end{bmatrix} \tag{5.118}$$

$$(\Omega_a)^T \mathbb{I}_c \frac{\partial \Omega_a}{\partial q} = \begin{bmatrix} h_1 & h_2 & h_3 \end{bmatrix} \tag{5.119}$$

where

$$h_1 = 0 \tag{5.120}$$

$$h_2 = -I_1(\dot{\psi}\dot{\phi}c_\theta - \dot{\psi}^2 s_\theta c_\theta) - I_2(\dot{\psi}\dot{\theta}s_\theta s_\phi c_\phi + \dot{\psi}^2 s_\theta c_\theta s_\phi^2)$$
$$- I_3(\dot{\psi}^2 s_\theta c_\theta c_\phi^2 - \dot{\psi}\dot{\theta}s_\theta s_\phi c_\phi) \tag{5.121}$$

$$h_3 = I_2(-\dot{\theta}^2 s_\phi c_\phi - \dot{\psi}\dot{\theta}c_\theta s_\phi^2 + \dot{\psi}\dot{\theta}c_\theta c_\phi^2 + \dot{\psi}^2 c_\theta^2 s_\phi c_\phi)$$
$$+ I_3(-\dot{\psi}^2 c_\theta^2 s_\phi c_\phi + \dot{\psi}\dot{\theta}c_\theta s_\phi^2 - \dot{\psi}\dot{\theta}c_\theta c_\phi^2 + \dot{\theta}^2 s_\phi c_\phi) \tag{5.122}$$

The Euler–Lagrange equation for the torques is

$$F(\tau) = \begin{bmatrix} \tau_1 \\ \tau_2 \\ \tau_3 \end{bmatrix} = \begin{bmatrix} d_1 - h_1 \\ d_2 - h_2 \\ d_3 - h_3 \end{bmatrix} \tag{5.123}$$

where

$$\tau_1 = \quad I_1(\ddot{\phi}s_\theta + \dot{\phi}\dot{\theta}c_\theta - \ddot{\psi}s_\theta^2 - 2\dot{\psi}\dot{\theta}s_\theta c_\theta)$$
$$+ I_2(\ddot{\theta}c_\theta s_\phi c_\phi - \dot{\theta}^2 s_\theta s_\phi c_\phi - \dot{\theta}\dot{\phi}c_\theta s_\phi^2 + \dot{\theta}\dot{\phi}c_\theta c_\phi^2$$
$$+ \ddot{\psi}c_\theta^2 s_\phi^2 - 2\dot{\psi}\dot{\theta}s_\theta c_\theta s_\phi^2 + 2\dot{\psi}\dot{\phi}c_\theta^2 s_\phi c_\phi)$$
$$+ I_3(\ddot{\psi}c_\theta^2 c_\phi^2 - 2\dot{\psi}\dot{\theta}s_\theta c_\theta c_\phi^2 - 2\dot{\psi}\dot{\phi}c_\theta^2 s_\phi c_\phi$$
$$- \ddot{\theta}c_\theta s_\phi c_\phi + \dot{\theta}^2 s_\theta s_\phi c_\phi + \dot{\theta}\dot{\phi}c_\theta s_\phi^2 - \dot{\theta}\dot{\phi}c_\theta c_\phi^2) \tag{5.124}$$

$$\tau_2 = I_1(\dot{\psi}\dot{\phi}c_\theta - \dot{\psi}^2 s_\theta c_\theta)$$
$$+ I_2(\ddot{\theta}c_\phi^2 - 2\dot{\theta}\dot{\phi}s_\phi c_\phi + \ddot{\psi}c_\theta s_\phi c_\phi + \dot{\psi}\dot{\phi}c_\theta c_\phi^2$$
$$- \dot{\psi}\dot{\phi}c_\theta s_\phi^2 + \dot{\psi}^2 s_\theta c_\theta s_\phi^2)$$
$$- I_3(\ddot{\psi}c_\theta s_\phi c_\phi + \dot{\theta}\dot{\phi}c_\theta c_\phi^2 - \dot{\psi}\dot{\phi}c_\theta s_\phi^2$$
$$- \ddot{\theta}s_\phi^2 - 2\dot{\theta}\dot{\phi}s_\phi c_\phi - \dot{\psi}^2 s_\theta c_\theta c_\phi^2) \tag{5.125}$$

$$\tau_3 = I_1(\ddot{\phi} - \ddot{\psi}s_\theta - \dot{\psi}\dot{\theta}c_\theta)$$
$$- I_2(-\dot{\theta}^2 s_\phi c_\phi - \dot{\psi}\dot{\theta}c_\theta s_\phi^2 + \dot{\psi}\dot{\theta}c_\theta c_\phi^2 + \dot{\psi}^2 c_\theta^2 s_\phi c_\phi)$$
$$- I_3(-\dot{\psi}^2 c_\theta^2 s_\phi c_\phi + \dot{\psi}\dot{\theta}c_\theta s_\phi^2 - \dot{\psi}\dot{\theta}c_\theta c_\phi^2 + \dot{\theta}^2 s_\phi c_\phi) \tag{5.126}$$

The nonlinear model, for the η coordinates, can be written as

$$M(\eta)\,\ddot{\eta} + C(\eta,\dot{\eta})\,\dot{\eta} = F(\tau) \tag{5.127}$$

where

$$M(\eta) = \begin{bmatrix} I_1 s_\theta^2 + I_2 c_\theta^2 s_\phi^2 + I_3 c_\theta^2 c_\phi^2 & I_2 c_\theta s_\phi c_\phi - I_3 c_\theta s_\phi c_\phi & -I_1 s_\theta \\ I_2 c_\theta s_\phi c_\phi - I_3 c_\theta s_\phi c_\phi & I_2 c_\phi^2 + I_3 s_\phi^2 & 0 \\ -I_1 s_\theta & 0 & I_1 \end{bmatrix} \tag{5.128}$$

$$C(\eta,\dot{\eta}) = \begin{bmatrix} c_{11} & c_{12} & c_{13} \\ c_{21} & c_{22} & c_{23} \\ c_{31} & c_{32} & c_{33} \end{bmatrix} \tag{5.129}$$

and

$$
\begin{aligned}
c_{11} = {}& I_1 \dot{\theta} s_\theta c_\theta - I_2 \dot{\theta} s_\theta c_\theta s_\phi^2 + I_2 \dot{\phi} c_\theta^2 s_\phi c_\phi \\
& - I_3 \dot{\theta} s_\theta c_\theta c_\phi^2 - I_3 \dot{\phi} c_\theta^2 s_\phi c_\phi
\end{aligned} \tag{5.130}
$$

$$
\begin{aligned}
c_{12} = {}& I_1 \dot{\psi} s_\theta c_\theta - I_2 \dot{\psi} s_\theta c_\theta s_\phi^2 - I_2 \dot{\theta} s_\theta s_\phi c_\phi - I_2 \dot{\phi} c_\theta s_\phi^2 + I_2 \dot{\phi} c_\theta c_\phi^2 \\
& - I_3 \dot{\psi} s_\theta c_\theta c_\phi^2 + I_3 \dot{\theta} s_\theta s_\phi c_\phi + I_3 \dot{\phi} c_\theta s_\phi^2 - I_3 \dot{\phi} c_\theta c_\phi^2
\end{aligned} \tag{5.131}
$$

$$c_{13} = -I_1 \dot{\theta} c_\theta + I_2 \dot{\psi} c_\theta^2 s_\phi c_\phi - I_3 \dot{\psi} c_\theta^2 s_\phi c_\phi \tag{5.132}$$

$$c_{21} = -I_1 \dot{\psi} s_\theta c_\theta + I_2 \dot{\psi} s_\theta c_\theta s_\phi^2 + I_3 \dot{\psi} s_\theta c_\theta c_\phi^2 \tag{5.133}$$

$$c_{22} = -I_2 \dot{\phi} s_\phi c_\phi + I_3 \dot{\phi} s_\phi c_\phi \tag{5.134}$$

$$
\begin{aligned}
c_{23} = {}& I_1 \dot{\psi} c_\theta - I_2 \dot{\psi} c_\theta s_\phi^2 + I_2 \dot{\psi} c_\theta c_\phi^2 - I_2 \dot{\theta} s_\phi c_\phi \\
& + I_3 \dot{\theta} s_\phi c_\phi + I_3 \dot{\psi} c_\theta s_\phi^2 - I_3 \dot{\psi} c_\theta c_\phi^2
\end{aligned} \tag{5.135}
$$

$$c_{31} = -I_2 \dot{\psi} c_\theta^2 s_\phi c_\phi + I_3 \dot{\psi} c_\theta^2 s_\phi c_\phi \tag{5.136}$$

$$
\begin{aligned}
c_{32} = {}& -I_1 \dot{\psi} c_\theta + I_2 \dot{\psi} c_\theta s_\phi^2 - I_2 \dot{\psi} c_\theta c_\phi^2 + I_2 \dot{\theta} s_\phi c_\phi \tag{5.137} \\
& - I_3 \dot{\theta} s_\phi c_\phi - I_3 \dot{\psi} c_\theta s_\phi^2 + I_3 \dot{\psi} c_\theta c_\phi^2 \tag{5.138}
\end{aligned}
$$

$$c_{33} = 0 \tag{5.139}$$

The left side of equation (5.107) is the total force applied to the helicopter (standard, tandem or coaxial). The right side of equation (5.127) represents the total torque applied to the helicopter (standard, tandem or coaxial).

In the case of the η coordinates, note that M is positive definite and

$$
\begin{aligned}
\dot{M}_{11} = {}& 2I_1 \dot{\theta} s_\theta c_\theta - 2I_2 \dot{\theta} s_\theta c_\theta s_\phi^2 + 2I_2 \dot{\phi} c_\theta^2 s_\phi c_\phi \\
& - 2I_3 \dot{\theta} s_\theta c_\theta c_\phi^2 - 2I_3 \dot{\phi} c_\theta^2 s_\phi c_\phi
\end{aligned} \tag{5.140}
$$

$$
\begin{aligned}
\dot{M}_{12} = {}& -I_2 \dot{\theta} s_\theta s_\phi c_\phi + I_2 \dot{\phi} c_\theta c_\phi^2 - I_2 \dot{\phi} c_\theta s_\phi^2 \\
& + I_3 \dot{\theta} s_\theta s_\phi c_\phi - I_3 \dot{\phi} c_\theta c_\phi^2 + I_3 \dot{\phi} c_\theta s_\phi^2
\end{aligned} \tag{5.141}
$$

$$\dot{M}_{13} = -I_1 \dot{\theta} c_\theta \tag{5.142}$$

$$\dot{M}_{21} = - I_2\dot{\theta}s_\theta s_\phi c_\phi + I_2\dot{\phi}c_\theta c_\phi^2 - I_2\dot{\phi}c_\theta s_\phi^2$$
$$+ I_3\dot{\theta}s_\theta s_\phi c_\phi - I_3\dot{\phi}c_\theta c_\phi^2 + I_3\dot{\phi}c_\theta s_\phi^2 \tag{5.143}$$

$$\dot{M}_{22} = - 2I_2\dot{\phi}s_\phi c_\phi + 2I_3\dot{\phi}s_\phi c_\phi \tag{5.144}$$

$$\dot{M}_{23} = 0 \tag{5.145}$$

$$\dot{M}_{31} = - I_1\dot{\theta}c_\theta \tag{5.146}$$

$$\dot{M}_{32} = 0 \tag{5.147}$$

$$\dot{M}_{33} = 0 \tag{5.148}$$

Moreover, the matrix

$$P = \dot{M} - 2C \tag{5.149}$$

is given by

$$P_{11} = 0 \tag{5.150}$$

$$P_{12} = - 2I_1\dot{\psi}s_\theta c_\theta + 2I_2\dot{\psi}s_\theta c_\theta s_\phi^2 + I_2\dot{\theta}s_\theta s_\phi c_\phi + I_2\dot{\phi}c_\theta s_\phi^2 - I_2\dot{\phi}c_\theta c_\phi^2$$
$$+ 2I_3\dot{\psi}s_\theta c_\theta c_\phi^2 - I_3\dot{\theta}s_\theta s_\phi c_\phi + I_3\dot{\phi}c_\theta c_\phi^2 - I_3\dot{\phi}c_\theta s_\phi^2 \tag{5.151}$$

$$P_{13} = I_1\dot{\theta}c_\theta - 2I_2\dot{\psi}c_\theta^2 s_\phi c_\phi + 2I_3\dot{\psi}c_\theta^2 s_\phi c_\phi \tag{5.152}$$

$$P_{21} = 2I_1\dot{\psi}s_\theta c_\theta - 2I_2\dot{\psi}s_\theta c_\theta s_\phi^2 - I_2\dot{\theta}s_\theta s_\phi c_\phi - I_2\dot{\phi}c_\theta s_\phi^2 + I_2\dot{\phi}c_\theta c_\phi^2$$
$$- 2I_3\dot{\psi}s_\theta c_\theta c_\phi^2 + I_3\dot{\theta}s_\theta s_\phi c_\phi - I_3\dot{\phi}c_\theta c_\phi^2 + I_3\dot{\phi}c_\theta s_\phi^2 \tag{5.153}$$

$$P_{22} = 0 \tag{5.154}$$

$$P_{23} = - 2I_1\dot{\psi}c_\theta + 2I_2\dot{\psi}c_\theta s_\phi^2 - 2I_2\dot{\psi}c_\theta c_\phi^2 + 2I_2\dot{\theta}s_\phi c_\phi$$
$$- 2I_3\dot{\theta}s_\phi c_\phi - 2I_3\dot{\psi}c_\theta s_\phi^2 + 2I_3\dot{\psi}c_\theta c_\phi^2 \tag{5.155}$$

$$P_{31} = - I_1\dot{\theta}c_\theta + 2I_2\dot{\psi}c_\theta^2 s_\phi c_\phi - 2I_3\dot{\psi}c_\theta^2 s_\phi c_\phi \tag{5.156}$$

$$P_{32} = 2I_1\dot{\psi}c_\theta - 2I_2\dot{\psi}c_\theta s_\phi^2 + 2I_2\dot{\psi}c_\theta c_\phi^2 - 2I_2\dot{\theta}s_\phi c_\phi$$
$$+ 2I_3\dot{\theta}s_\phi c_\phi + 2I_3\dot{\psi}c_\theta s_\phi^2 - 2I_3\dot{\psi}c_\theta c_\phi^2 \tag{5.157}$$

$$P_{33} = 0 \tag{5.158}$$

which is skew-symmetric, and this result proves that the system of the η coordinates is passive. The translational part cannot be proved to be passive. This is due to the effect of the potential energy of gravitation that does not satisfy the requirements of a dissipative system [94].

5.4 Nonlinear Control Strategy

In this section, we propose a controller based on backstepping techniques for both Newton–Euler and Euler–Lagrange dynamic models.

Newton–Euler dynamic model

The block diagram of Figure 5.7 represents an input–output system of equations (5.73)–(5.76), where u and γ are the inputs, and the output is the position ξ. Note that the coupling terms $\mathcal{R}\sigma\gamma$ and $k_0 u$ result in feed-forward and feedback connections, which destroy the pure cascade structure of the system. This cascade structure is needed in order to use the backstepping procedure. Thus, consider that the small-body forces have a value lower than the lift force u, valid in hover conditions. Consider also that eccentricity between the rotor hub and the E_3 axis ($k_0 = 0$) does not exist. With these assumptions, the helicopter dynamic model has a cascade structure. We will show, in simulations, that these assumptions do not alter the control objectives.

We wish to fully determine the trajectories of the helicopter. The position trajectory is a simple matter to assign. To achieve a given trajectory, it will be necessary to manipulate the direction in which the principal translation force u acts and this will in turn determine trajectories for the pitch and roll of the helicopter. The yaw, however, is also a free variable, so we will assign a trajectory to this variable.

Let

$$\xi^d : \mathbb{R} \to \mathbb{R}^3$$
$$\psi^d : \mathbb{R} \to \mathbb{R}$$

be smooth trajectories $\xi^d(t) = (x^d(t), y^d(t), z^d(t))$ and $\psi^d(t)$. The control problem considered is to find a feedback control action $(u, \gamma_1, \gamma_2, \gamma_3)$ depending only on the measurable states $(\xi, \dot{\xi}, \eta, \dot{\eta})$ and arbitrarily many derivatives of the smooth trajectory $(\xi^d(t), \psi^d(t))$ such that the tracking error

$$\mathcal{E} := (\xi(t) - \xi^d(t), \psi(t) - \psi^d(t)) \in \mathbb{R}^4 \tag{5.159}$$

is asymptotically stable for all initial conditions.

Define a partial error

$$\delta_1 = \xi - \xi^d \tag{5.160}$$

and the first storage function as

$$S_1 = \frac{1}{2}\delta_1^T \delta_1 = \frac{1}{2}|\delta_1|^2 \tag{5.161}$$

The time derivative of S_1 is given by

$$\dot{S}_1 = \delta_1^T \dot{\delta}_1 = \delta_1^T (v - v^d) \tag{5.162}$$

where v^d represents the velocity of the tracked trajectory. Let v^v be the virtual control for the first stage of this procedure and be chosen as

$$v^v = v^d - k_1\delta_1 \tag{5.163}$$

where k_1 is a positive constant. We will use these positive constants (k_2, k_3, etc.) in the controller design.

Introducing the above into (5.162), we get

$$\dot{S}_1 = -k_1|\delta_1|^2 + \frac{1}{m}\delta_1^T\delta_2 \tag{5.164}$$

where δ_2 is defined as

$$\delta_2 = mv - mv^v \tag{5.165}$$

Differentiating δ_2 and using equation (5.74) without the small-body forces, it follows that

$$\dot{\delta}_2 = -u\mathcal{R}E_3 + mgE_z - m\dot{v}^v \tag{5.166}$$

Define a second storage function associated with the second error term δ_2

$$S_2 = \frac{1}{2}|\delta_2|^2 \tag{5.167}$$

Therefore, from (5.166)

$$\dot{S}_2 = \delta_2^T(-u\mathcal{R}E_3 + mgE_z - m\dot{v}^v) \tag{5.168}$$

Consider the new virtual control:

$$X^v = (u\mathcal{R}E_3)^v = mgE_z - m\dot{v}^v + \frac{1}{m}\delta_1 + k_2\delta_2 \tag{5.169}$$

In this case, the derivative of the second storage function becomes

$$\dot{S}_2 = -\frac{1}{m}\delta_2^T\delta_1 - k_2|\delta_2|^2 + \delta_2^T\delta_3 \tag{5.170}$$

where δ_3 is the third error used in this procedure. It is defined as

$$\delta_3 = X^v - u\mathcal{R}E_3 \tag{5.171}$$

Consider the new storage function

$$S_3 = \frac{1}{2}|\delta_3|^2 + \frac{1}{2}|\epsilon_3|^2 \tag{5.172}$$

where

$$\epsilon_3 = \psi - \psi^d \tag{5.173}$$

penalizes the yaw. The yaw component of the error term is introduced at this stage of the backstepping procedure in order that the relative degrees of δ_3 and ϵ_3 with respect to the controls u and γ match. Indeed, the relative degree of each control with respect to either error is two. Differentiating the above expression and recalling (5.75), we have

$$\dot{S}_3 = \delta_3^T \dot{\delta}_3 + \epsilon_3(\dot{\psi} - \dot{\psi}^d) \tag{5.174}$$

or using (5.30), (5.75) and (5.171)

$$\dot{S}_3 = \delta_3^T(\dot{X}^v - \dot{u}RE_3 - uR\hat{\Omega}E_3) + \epsilon_3(\dot{\psi} - \dot{\psi}^d)$$
$$= \delta_3^T \left(\dot{X}^v - R \begin{bmatrix} u\Omega_2 \\ -u\Omega_1 \\ \dot{u} \end{bmatrix} \right) + \epsilon_3(\dot{\psi} - \dot{\psi}^d) \tag{5.175}$$

The value of $\dot{u}$ can be assigned directly via the following control law:

$$\dot{u} = E_3^T R^T(\dot{X}^v + \delta_2 + k_3\delta_3) \tag{5.176}$$

Note that if we assume that the measurements of $(\xi, \dot{\xi}, \eta, u)$ are available, then we can estimate the value of the derivative of the thrust $\dot{u}$. When the measurement of u is not available, but the measurement of $\ddot{\xi}$ is available, then we can estimate the value of u, using the relation

$$|u| = m\left|\ddot{\xi} - gE_z\right| \tag{5.177}$$

Now, define the following virtual input:

$$\begin{pmatrix} u\Omega_2 \\ -u\Omega_1 \\ 0 \end{pmatrix}^v = [I - E_3E_3^T](\dot{X}^v + \delta_2 + k_3\delta_3) = Y^v \tag{5.178}$$

To proceed we introduce the error variable

$$\delta_4 = Y^v - uR\hat{\Omega}E_3 \tag{5.179}$$

Thus, equation (5.175) becomes

$$\dot{S}_3 = -\delta_3^T \delta_2 - k_3|\delta_3|^2 + \delta_3^T \delta_4 + \epsilon_3(\dot{\psi} - \dot{\psi}^d) \tag{5.180}$$

Now consider the term associated with ϵ_3. Let $\dot{\psi}^v$ denote the virtual yaw velocity and choose

$$\dot{\psi}^v = \dot{\psi}^d - k_{31}\epsilon_3 \tag{5.181}$$

We can rewrite (5.180) as

$$\dot{S}_3 = -\delta_3^T \delta_2 - k_3|\delta_3|^2 + \delta_3^T \delta_4 - k_{31}\epsilon_3^2 + \epsilon_3(\dot{\psi} - \dot{\psi}^v) \tag{5.182}$$

Define $\epsilon_4 = \dot{\psi} - \dot{\psi}^v$. With this choice the derivative of S_3 is

$$\dot{S}_3 = -\delta_3^T \delta_2 - k_3|\delta_3|^2 + \delta_3^T \delta_4 - k_{31}\epsilon_3^2 + \epsilon_3\epsilon_4 \tag{5.183}$$

The fourth storage function associated with the backstepping procedure is given by

$$S_4 = \frac{1}{2}|\delta_4|^2 + \frac{1}{2}|\epsilon_4|^2 \tag{5.184}$$

Taking the derivative of S_4 yields

$$\dot{S}_4 = \delta_4^T(\dot{Y}^v - (\tilde{u}\mathcal{R}E_3 - u\mathcal{R}\hat{E}_3\tilde{\gamma})) + \epsilon_4(\ddot{\psi} - \ddot{\psi}^v) \tag{5.185}$$

where $\tilde{\gamma} = \dot{\Omega}$ is a control input transformation. Now to achieve the desired control, we choose

$$\tilde{u}\mathcal{R}E_3 - u\mathcal{R}\hat{E}_3\tilde{\gamma} = \dot{Y}^v + \delta_3 + k_5\delta_4 \tag{5.186}$$

$$\ddot{\psi} = \ddot{\psi}^v - \epsilon_3 - k_6\epsilon_4 \tag{5.187}$$

Introducing the above equations into (5.185) we get

$$\dot{S}_4 = -\delta_4^T\delta_3 - k_4|\delta_4|^2 - k_{41}|\epsilon_4|^2 - \epsilon_4\epsilon_3 \tag{5.188}$$

It remains to define the control signal equations as

$$\tilde{u} = E_3^T\mathcal{R}^T\left(\dot{X}^v + \delta_2 + k_3\delta_3\right) \tag{5.189}$$

$$\tilde{\gamma}^1 = -\frac{E_2^T\mathcal{R}^T}{u}\left(\dot{Y}^v - \tilde{u}\Omega^1 + \delta_3 + k_5\delta_4\right) \tag{5.190}$$

$$\tilde{\gamma}^2 = \frac{E_1^T\mathcal{R}^T}{u}\left(\dot{Y}^v - \tilde{u}\Omega^2 + \delta_3 + k_5\delta_4\right) \tag{5.191}$$

$$\tilde{\gamma}^3 = \frac{\cos\theta}{\cos\phi}\left(\ddot{\psi}^v - \epsilon_3 - k_6\epsilon_4 + E_1^TW_\eta^{-1}\dot{W}_\eta W_\eta^{-1}\Omega - \frac{\sin\phi}{\cos\theta}\tilde{\gamma}^2\right) \tag{5.192}$$

where $\tilde{u} = \dot{u}$. The equation of $\tilde{\gamma}^3$ was obtained from the second derivative of η. This is due to the absence of this variable from the control equation design. Then, we have

$$\dot{\eta} = \frac{1}{\cos\theta}\begin{pmatrix} 0 & \sin\phi & \cos\phi \\ 0 & \cos\theta\cos\phi & -\cos\theta\sin\phi \\ \cos\theta\sin\theta\sin\phi & \sin\theta\cos\phi \end{pmatrix}\Omega$$

$$= W_\eta^{-1}\Omega \tag{5.193}$$

where

$$W_\eta = \begin{pmatrix} -\sin\theta & 0 & 1 \\ \cos\theta\sin\phi & \cos\phi & 0 \\ \cos\theta\cos\phi & -\sin\phi & 0 \end{pmatrix} \tag{5.194}$$

and

$$\ddot{\eta} = -W_\eta^{-1}\dot{W}_\eta W_\eta^{-1}\Omega + W_\eta^{-1}\dot{\Omega}$$

$$= -W_\eta^{-1}\dot{W}_\eta W_\eta^{-1}\Omega + W_\eta^{-1}\tilde{\gamma} \tag{5.195}$$

Obtaining $\ddot{\psi}$ yields

$$\ddot{\psi} = -E_1^T W_\eta^{-1} \dot{W}_\eta W_\eta^{-1} \Omega + \frac{\sin\phi}{\cos\theta}\tilde{\gamma}^2 + \frac{\cos\phi}{\cos\theta}\tilde{\gamma}^3 \qquad (5.196)$$

The backstepping process shown before accomplishes the monotonic decrease of the following Lyapunov function

$$V = S_1 + S_2 + S_3 + S_4$$
$$= \frac{1}{2}\sum_{i=1}^{4}|\delta_i|^2 + \frac{1}{2}|\epsilon_3|^2 + \frac{1}{2}|\epsilon_4|^2 \qquad (5.197)$$

One can directly verify that

$$\dot{V} = -k_1|\delta_1|^2 - k_2|\delta_2|^2 - k_3|\delta_3|^2 - k_4|\delta_4|^2 - k_{31}|\epsilon_3|^2 - k_{41}|\epsilon_4|^2 \quad (5.198)$$

Note that δ_1 and ϵ_3 together form the original tracking error that we wish to minimize. Then the Lyapunov function V is monotonically decreasing and thus the control objective is achieved.

A backstepping controller design for a dynamic model represented in a body-fixed frame can be found in [35].

Euler–Lagrange dynamic model

This controller was proposed by Mahony et al. in [97]. The problem considered is also that of smooth path tracking. In particular, we consider a given path in the coordinates $\xi = (x, y, z)$ and look for a control law that manipulates the full generalized coordinates to ensure that the path is followed. In addition to the path coordinates in ξ, we add additional trajectory requirements on the yaw angle $\psi(t)$ and on the regulation of the rotor speed $\dot{\gamma}$. The specified path is practically motivated by the desire to regulate the position, orientation and rotor speed of a helicopter in hover. The desired trajectories do not fit into the standard framework for backstepping path tracking designs.

Consider the model of a helicopter given by

$$m\ddot{\xi} + mgE_3 = F_\xi = uG(\eta) \qquad (5.199)$$
$$\mathbb{I}\ddot{\eta} + C(\eta, \dot{\eta})\dot{\eta} = \tau \qquad (5.200)$$

where $uG(\eta) = u\mathcal{R}E_3$ and $C(\eta, \dot{\eta})$ is defined by equation (5.129).

Let

$$\xi^d : \mathbb{R} \to \mathbb{R}^3$$
$$\psi^d : \mathbb{R} \to \mathbb{R}$$

be smooth trajectories $\xi^d(t) := (x^d(t), y^d(t), z^d(t))$ and $\psi^d(t)$. Let $\kappa > 0$ be a constant. The control problem considered is to find a feedback control action $(u, \tau) \in \mathbb{R}^4$ depending only on the measurable states $(\dot{\xi}, \xi, \dot{\eta}, \eta)$ and arbitrarily many derivatives of the smooth trajectory $(\xi^d(t), \psi^d(t))$ such that the tracking error

$$\mathcal{E} := (\xi(t) - \xi^d(t), \psi^d(t) - \psi^d(t), \dot{\gamma}(t) - \kappa) \in \mathcal{R}^5 \tag{5.201}$$

is asymptotically stable. $\dot{\gamma}$ represents the rotor angular speed.

Two points need to be emphasized regarding the control problem as stated. Firstly, the desired trajectory must be sufficiently smooth before the techniques employed in the sequel may be applied. We will be keeping in mind the problem of regulation to a set point for hover regulation as the prime example. Secondly, the tracking error as defined is a mixture of paths in the translation coordinates, the orientation coordinates and derivatives of the orientation coordinates. These errors all have different relative degrees with respect to the control inputs and preclude the direct application of standard backstepping techniques.

Consider once again (5.199) and (5.200). It is clear that these equations are in block pure feedback form [89], where the first block is (5.199). This leads one to consider a partial error in the variables $\xi = (x, y, z)$ and to use this to backstep, adding additional error variables as appropriate.

Consider the error

$$e := \xi(t) - \xi^d(t) \tag{5.202}$$

Then, following the standard approach for path tracking in mechanical systems [153], we consider the output

$$\alpha := \dot{e} + e \tag{5.203}$$

The choice of α is motivated by a zero dynamics argument. That is, if one designs a controller to drive $\alpha \to 0$, then the zero dynamics

$$\dot{\alpha} = -\alpha \tag{5.204}$$

are globally and asymptotically stable and ensure that the error itself will also converge to zero. Taking the time derivative of α and substituting for (5.199) yields

$$\begin{aligned} m\frac{d}{dt}\alpha &= m\ddot{e} + m\dot{e} \\ &= m(\dot{e} - \ddot{\xi}^d) + m\ddot{\xi} \\ &= m(\dot{e} - \ddot{\xi}^d) - mgE_z + G(\eta)u \end{aligned} \tag{5.205}$$

Formally, there is only a single input "u" present in this equation and it is impossible to assign the desired three-dimensional stable dynamics. The process of backstepping suggests that we consider the variables η as inputs themselves. In this case, it is clear that the unit vector $G(\eta)$ may be arbitrarily assigned direction and that the control u can be used to assign the magnitude desired for the stable dynamics required. Unfortunately, such an approach brings its own problems since formally there are now four input variables (η, u) to assign three-dimensional dynamics. Moreover, solving the vector $G(\eta)$ for the angles (ψ, θ, ϕ) introduces some unpleasant nonlinearities if the resulting explicit expressions are used in a backstepping design. An indication of the complications involved in an approach like this is present in the design of explicit backstepping control of a VTOL aircraft [147]. Rather than take this approach, we will view the vector $G(\eta)u$ as a vector in $\mathbb{R}^3$ and carry the full expression through to the backstepping procedure. Thus, we define an error

$$\beta_1 = G(\eta)u - X \tag{5.206}$$

where

$$X := X(\ddot{\xi}^d, \dot{\xi}^d, \xi^d, \dot{\xi}, \xi) \tag{5.207}$$

is a function of known signals and is chosen to assign stable dynamics to the error α. In particular, choose

$$X = -\left(m(\dot{e} - \ddot{\xi}^d) - mgE_z + \alpha \right) \tag{5.208}$$

Consider the storage function

$$S_\alpha = \frac{m}{2}|\alpha|^2 = \frac{m}{2}\alpha^T \alpha \tag{5.209}$$

Differentiating the storage S_α, one obtains

$$\dot{S}_\alpha = -|\alpha|^2 + \alpha^T \left(G(\eta)u - X \right)$$
$$= -|\alpha|^2 + \alpha^T \beta_1 \tag{5.210}$$

In the formal process of backstepping, the error β_1 would now be differentiated and stable dynamics assigned to it in turn. Such an approach requires that time derivatives of the input u are computed. This can be achieved by dynamically extending the input u so that it has the same relative degree in (5.199) as the variables ν. Thus, we rewrite the helicopter dynamics, (5.199) and (5.200), and augment these dynamics with a cascade of two integrators feeding into the control action u

$$m\ddot{\xi} = G(\eta)u - mgE_z \tag{5.211}$$
$$\dot{\eta} = \eta_2 \tag{5.212}$$
$$\dot{u} = u_2 \tag{5.213}$$
$$\dot{\eta}_2 = \ddot{\eta} = \mathbb{I}^{-1}\tau - \mathbb{I}^{-1}C(\eta, \dot{\eta})\dot{\eta} \tag{5.214}$$
$$\dot{u}_2 = \ddot{u} = v \tag{5.215}$$

where the new variable $v \in \mathbb{R}$ along with the original generalized torques τ are the inputs for the augmented system. Note that the v and τ inputs now both have a relative degree of four with respect to the ξ coordinates. Moreover, note that the error α has a relative degree of three with respect to the inputs (τ, v). The kinematic equations (5.212) and (5.213) are included to display explicitly the block pure feedback form of the equations. This occurs in three cascaded blocks, the first (5.211), the second (5.212) and (5.213), and the final block (5.214) and (5.215).

The design error β_1 is an error in the coordinates η and the control u. It is easily seen that β_1 has a relative degree of two with respect to the inputs (τ, v). Recalling the additional tracking errors given in $\mathcal{E}$, note that $(\psi - \psi^d)$ also has a relative degree of two while $(\dot{\gamma} - \kappa)$ has a relative degree of one. Thus, before continuing with the formal backstepping procedure, it is possible to augment β_1 with an additional term that accounts for the tracking performance of the yaw $(\psi - \phi^d)$. The final tracking error $(\dot{\gamma} - \kappa)$ will be saved until the last step of the design procedure. Consider the augmented design error

$$\beta = (\beta_1^T, (\psi - \psi^d))^T = \begin{pmatrix} G(\eta)u - X \\ \psi - \psi^d \end{pmatrix} \in \mathbb{R}^4 \qquad (5.216)$$

The design error $\beta := \beta_t(\nu, u)$ can be thought of as a time-varying function of (ν, u). It is a straightforward, though somewhat tedious, calculation to show that in a suitable neighbourhood of $(\nu, u) = (\psi, 0, 0, u_0)$ for non-zero u_0 then the error β is a diffeomorphism. That is, it maps the four variables (ψ, θ, ϕ, u) locally one-to-one into $\mathbb{R}^4$. Thus, the addition of the extra tracking condition $(\psi \to \psi^d)$ has removed the difficulty associated with uniquely defining a trajectory for the variables ν that arose when first defining β_1. Given that the map β is locally a diffeomorphism on the domain of interest, the Jacobian

$$J(\nu, u) := \frac{\partial \beta}{\partial(\nu, u)}(\nu, u) \in \mathbb{R}^{4 \times 4} \qquad (5.217)$$

is also well defined and non-singular in this domain.

Taking the time derivative of β yields

$$\frac{d}{dt}\beta = J(\nu, u) \begin{pmatrix} \dot{\nu} \\ \dot{u} \end{pmatrix} - \begin{pmatrix} \dot{X} \\ \dot{\psi}^d \end{pmatrix} \qquad (5.218)$$

Note that

$$
\begin{aligned}
\dot{X} &= -\dot{\alpha} - m\left(\ddot{e} - \frac{d\ddot{\xi}^d}{dt}\right) = m\frac{d\ddot{\xi}^d}{dt} - (1+m)\ddot{e} - \dot{e} \\
&= m\frac{d\ddot{\xi}^d}{dt} + (1+m)\ddot{\xi}^d - \frac{(1+m)}{m}(G(\eta)u - mgE_z) - \dot{e} \\
&:= \dot{X}\left(\frac{d\ddot{\xi}^d}{dt}, \ldots, \xi^d, \dot{\xi}, \xi, \eta, u\right)
\end{aligned}
\qquad (5.219)
$$

Thus, one can define $Y = Y(\frac{d\ddot{\xi}^d}{dt}, \ldots, \xi^d, \dot{\xi}, \xi, \eta, u)$ by

$$Y := \begin{pmatrix} \dot{X} \\ \dot{\psi}^d \end{pmatrix} - \beta - \begin{pmatrix} \alpha \\ 0 \end{pmatrix} \in \mathbb{R}^4 \tag{5.220}$$

Observe that

$$\beta^T \begin{pmatrix} \alpha \\ 0 \end{pmatrix} = \alpha^T \beta_1 \tag{5.221}$$

since the fourth coordinate of β is cancelled by a zero. Consider the storage function

$$S_\beta := \frac{1}{2}|\beta|^2 \tag{5.222}$$

then it is easily verified that

$$\dot{S}_\beta = -|\beta|^2 - \alpha^T \beta_1 + \beta^T \left(J(\nu, u) \begin{pmatrix} \dot{\nu} \\ \dot{u} \end{pmatrix} - Y \right) \tag{5.223}$$

Following the backstepping methodology, a third design error is defined

$$\delta_1 := \left(J(\nu, u) \begin{pmatrix} \dot{\nu} \\ \dot{u} \end{pmatrix} - Y \right) \tag{5.224}$$

The next step is to backstep once more with the error δ_1. However, before this is done, it is possible to augment the error δ_1 with an error to guarantee the final tracking requirement $\dot{\gamma} \to \kappa$. Once again, the tracking error for γ is introduced when its relative degree matches the relative degree of the design error propagated from an earlier block. Thus, let

$$\delta = \begin{pmatrix} \delta_1 \\ \dot{\gamma} - \kappa \end{pmatrix} \tag{5.225}$$

To simplify the following structure, define

$$\overline{J}(\eta, u) := \begin{pmatrix} J(\nu, u) & 0 \\ 0 & 1 \end{pmatrix}, \quad \overline{Y} := \begin{pmatrix} Y \\ \kappa \end{pmatrix} \tag{5.226}$$

then

$$\delta := \left(\overline{J}(\eta, u) S \begin{pmatrix} \dot{\eta} \\ \dot{u} \end{pmatrix} - \overline{Y} \right) \tag{5.227}$$

where $S \in \mathbb{R}^{5 \times 5}$ is a permutation matrix that exchanges entries four and five of $(\dot{\eta}, \dot{u})$, ensuring that the $\dot{\gamma}$ and $\dot{u}$ terms match up with the correct rows of $\overline{J}(\eta, u)$.

Consider the derivative of δ

$$\dot{\delta} = \overline{J}(\eta, u) S \begin{pmatrix} \ddot{\eta} \\ \ddot{u} \end{pmatrix} + \dot{\overline{J}}(\eta, u) S \begin{pmatrix} \dot{\eta} \\ \dot{u} \end{pmatrix} - \dot{\overline{Y}} \tag{5.228}$$

Though an explicit notation for the Jacobian $\overline{J}(\eta, u)$ is required so that the linear dependence of the second derivatives $(\ddot{\eta}, \ddot{u}, \dot{\hat{\gamma}})$ is clearly expressed, the remaining terms in the above expressions are left as simple time derivatives due to space restrictions. It should be noted, however, that the evaluation of $\dot{\overline{J}}(\eta, u)$ and $\dot{\overline{Y}}$ depends only on known signals.

Let $r > 0$ be a constant gain and define a vector $Z := Z(\hat{\xi}^{(4)}, \ldots, \hat{\xi}, \dot{\xi}, \xi, \dot{\eta}, \eta, \dot{u}, u)$ by

$$Z = -\dot{\overline{J}}(\eta, u)S\begin{pmatrix} \dot{\eta} \\ \dot{u} \end{pmatrix} + \dot{\overline{Y}} - r\delta - \begin{pmatrix} \beta \\ 0 \end{pmatrix} \in \mathbb{R}^5 \tag{5.229}$$

The gain $r > 0$ is introduced to assign a rate of convergence to the δ error dynamics that are independent of the rates of convergence chosen for the other error variables. The reason for this choice is related to a robustness analysis [43]. Consider a third storage function

$$S_\delta = \frac{1}{2}|\delta|^2 \tag{5.230}$$

Computing the time derivative of S_δ yields

$$\frac{d}{dt}S_\delta = -r|\delta|^2 - \beta^T \delta_1 + \delta^T \left(\overline{J}(\eta, u)S\begin{pmatrix} \ddot{\eta} \\ \ddot{u} \end{pmatrix} - Z \right) \tag{5.231}$$

Define

$$K(\eta, u) := \overline{J}(\eta, u)S\begin{pmatrix} \mathbb{I}^{-1} & 0 \\ 0 & 1 \end{pmatrix} \in \mathbb{R}^{5\times 5} \tag{5.232}$$

It follows directly from the invertibility of $J(\eta, u)$, S and $\mathbb{I}$ that $K(\eta, u)$ is invertible. Thus, using (5.214) and (5.215), one can write

$$\overline{J}(\eta, u)S\begin{pmatrix} \ddot{\eta} \\ \ddot{u} \end{pmatrix} = K(\eta, u)\begin{pmatrix} \tau \\ v \end{pmatrix} + \overline{J}(\eta, u)S\begin{pmatrix} -\mathbb{I}^{-1}C(\eta, \dot{\eta})\dot{\eta} \\ 0 \end{pmatrix} \tag{5.233}$$

Since the matrix $K(\eta, u)$ is full rank and the vector (τ, v) is a full rank vector of inputs, then the nonlinear contribution to the δ dynamics may be explicitly cancelled by choosing

$$\begin{pmatrix} \tau \\ v \end{pmatrix} := K(\eta, u)^{-1}\overline{J}(\eta, u)S\begin{pmatrix} \mathbb{I}^{-1}C(\eta, \dot{\eta})\dot{\eta} \\ 0 \end{pmatrix} + K(\eta, u)^{-1}Z \tag{5.234}$$

As a consequence of applying this control, one has

$$\dot{S}_\delta = -r|\delta|^2 - \beta^T \delta_1 \tag{5.235}$$

Proposition 5.2. *Consider the augmented dynamics of a helicopter given by (5.211)–(5.215) and assume a desired trajectory $(\hat{\xi}, \hat{\phi}, \kappa)$ is given according to the definition of $\mathcal{E}$. Then, if the closed-loop trajectory evolves such that Euler angle representation of the airframe orientation remains well defined for all time t and the applied thrust u is never zero, then the control law (τ, v) given by (5.234) ensures exponential stabilization of the tracking error $\mathcal{E}$ in (5.201).*

Proof.   Consider the combined storage function

$$S = \frac{1}{2}(|e|^2 + S_\alpha + S_\beta + S_\delta) \tag{5.236}$$

Recalling the definition of α, one has

$$\dot{e} = -e + \alpha \tag{5.237}$$

Thus

$$\frac{1}{2}\frac{d}{dt}|e|^2 = -|e|^2 + e^T\alpha$$
$$= -\frac{1}{2}|e|^2 - \frac{1}{2}|e-\alpha|^2 + \frac{1}{2}|\alpha|^2 \tag{5.238}$$

The time derivative of S is computed by substituting the above calculation along with the expressions obtained earlier for the derivatives of S_α, S_β and S_δ.

$$\dot{S} = -\frac{1}{2}|e|^2 - \frac{1}{2}|e-\alpha|^2 + \frac{1}{2}|\alpha|^2 - |\alpha|^2 + \alpha^T\beta_1 - |\beta|^2 - \alpha^T\beta_1$$
$$+ \beta^T\delta_1 - r|\delta|^2 - \beta^T\delta_1$$
$$= -\frac{1}{2}|e|^2 - \frac{1}{2}|e-\alpha|^2 - \frac{1}{2}|\alpha|^2 - |\beta|^2 - r|\delta|^2 \tag{5.239}$$

It follows that $\dot{S} < 0$ unless $e = 0 = \alpha = \beta = \delta$. Thus, by applying Lyapunov's theorem, all the errors e, α, β and δ are asymptotically stable to zero and the desired tracking is achieved. $\square$

5.5 Simulations

In this section, we present the simulation of the behaviour of the complete helicopter dynamics and the approximative dynamics used to obtain the control law for the case of the Newtonian model. The experiment considers the case of stabilization of the standard helicopter dynamics to a stationary configuration. The parameters used for the standard helicopter model are based on a VARIO 23 cc small helicopter (see Figure 5.8) and are given in Table 5.1. The initial and desired positions are defined as

$$\xi_0 = \begin{pmatrix} 0 \\ 4 \\ -5 \end{pmatrix} m, \quad \psi_0 = 0° \tag{5.240}$$

$$\xi^d = \begin{pmatrix} 2 \\ 6 \\ -15 \end{pmatrix} m, \quad \psi^d = 60° \tag{5.241}$$

Fig. 5.8. VARIO 23 cc small helicopter.

The initial condition adopted for the control is $u = gm \approx 74$. It is exactly equal to the force required for sustaining the helicopter in stationary flight. The values of the gains are: $k_1 = 0.25$, $k_2 = 1$, $k_3 = 1$, $k_4 = 10$, $k_{31} = 0.1$ and $k_{41} = 6$.

Parameter	Value
m	7.5 kg
I_{11}	0.177 kg m^2
I_{12}	$-$ 0.008 kg m^2
I_{21}	$-$ 0.008 kg m^2
I_{22}	0.698 kg m^2
I_{33}	0.704 kg m^2
$|Q_M|$	0.02 $|T_M|$
$|Q_T|$	0.02 $|T_T|$
l_M	1.2 m
l_T	0.27 m
g	9.80 m s^{-2}

Table 5.1. Physical parameters of the VARIO 23 cc small helicopter.

Figures 5.9 to 5.11 illustrate the system behaviour since the adapted dynamic model is used ($\sigma = 0$). The behaviour of the position (ξ) and the orientation (η) is shown in Figure 5.9; we can verify that the position and orientation desired are achieved. Figure 5.10 shows the decrease of the Lyapunov function and its components.

The cyclic angles of the main rotor (a and b that generate the translational displacement $x - y$) and the magnitudes of the thrust vectors (T_M and T_T) are showed in Figure 5.11. We can verify that the value of angle a does not converge to zero because the tail rotor produces a torque in yaw that is compensated by this tilt.

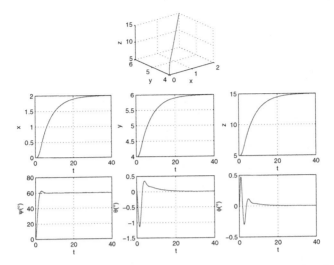

Fig. 5.9. Position regulation of the helicopter dynamics in the absence of small-body forces.

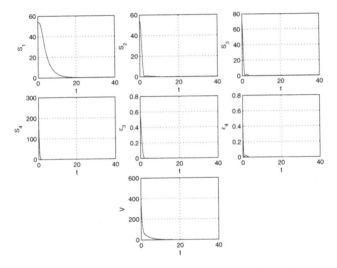

Fig. 5.10. The Lyapunov function and its components for position regulation in the absence of small-body forces.

The variation of the value of T_T denotes the change of force needed to obtain the desired orientation of the angle (ψ^d). The value of T_M incites the translation movement in the E_z axis. On the other hand, Figures 5.12 to 5.14 illustrate the system behaviour in the presence of small-body forces. The position and orientation variables are showed in Figure 5.12.

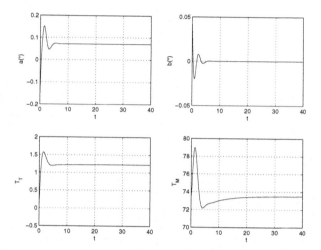

Fig. 5.11. Cyclic angles and thrust vectors of the helicopter dynamics in the absence of small-body forces.

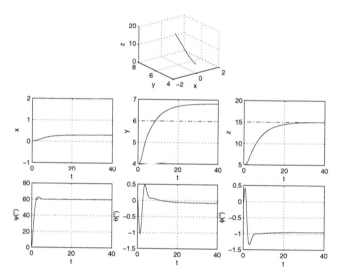

Fig. 5.12. Position regulation of the helicopter dynamics in the presence of small-body forces.

We can see that the system is stable but the position is not achieved for the x and y coordinates. Analysing the stability of the system [36], we deduce that one of the conditions for the system convergence is attached to the anti-torques $\Gamma_Q = 0$. This condition is not satisfied (Table 5.1).

Figure 5.13 shows the decrease of the Lyapunov function and its components (storage functions). Note that these values do not converge completely to zero; this is due to the consideration of the full dynamic model (k_0, σ, etc.). However, the function values are stabilized to constant values (bounded).

The variation of the cyclic angles and the amplitudes of the thrust vectors are presented in Figure 5.14. Figure 5.15 illustrates the parametric robustness of the control law with respect to cyclic angles of the main rotor and amplitudes of the thrust vectors. The values considered in this test are $\mathbf{I} = 1.25\mathbf{I}$ and $\mathbf{I} = 0.75\mathbf{I}$ where the parameters of the inertial matrix are given in Table 5.1.

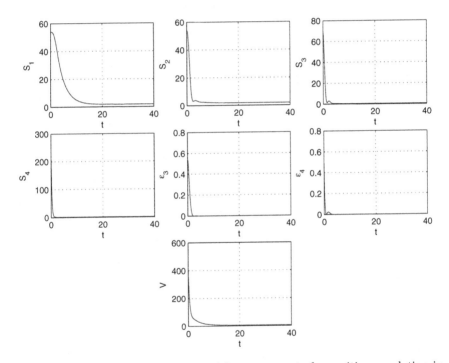

Fig. 5.13. The Lyapunov function and its components for position regulation in the presence of small-body forces.

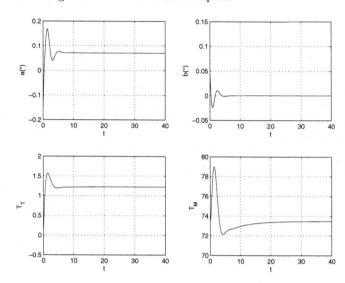

Fig. 5.14. Cyclic angles and thrust vectors of the helicopter dynamics in the presence of small-body forces.

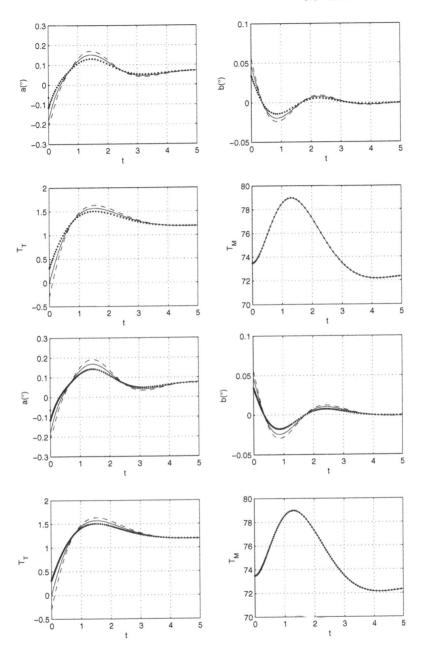

Fig. 5.15. Behaviour of the cyclic angles and thrust vectors. Solid for **I**, dashed for 1.25**I** and dotted for 0.75**I**, in the absence of the small-body forces for the upper side, and in the presence of the small-body forces for the under side.

6

Helicopter in a Vertical Flying Stand

6.1 Introduction

Practical flight control problems for small unmanned helicopters becomes a challenge for the control community. Modelling and parameter identification, for these flying machines, are essential for obtaining control algorithms with satisfactory performance [169]. Tischler et al. [175] introduced a technique called Comprehensive Identification from Frequency Responses (CIFER) to identify the parameters of a 9 degree-of-freedom (DoF) hybrid model of a full-scale BO 105 dynamics from flight test data at 80 knots. Mettler et al. [104] applied the CIFER identification technique to a Yamaha R-50 small-scale helicopter. They proposed and validated two linear models for hover and cruise flight. Recently La Civita et al. [90] introduced a novel modelling technique called Modelling for Flight Simulation and Control Analysis (MOSCA). This method has been applied to identify the model parameters of the Carnegie Mellon University Yamaha R-50 helicopter. Gavrilets et al. [56] developed and validated a nonlinear model for small-size helicopters using flight data collected on an X-Cell-60 acrobatic helicopter. The model has been validated for a wide range of conditions including acrobatic manoeuvres. Weilenmann et al. [184] investigated a small-size helicopter mounted on a mechanical platform allowing 6 degree of freedom. They assumed the system to be represented by a linear model and applied different controller designs showing the superiority of modern multivariable concepts over classical approaches. Other identification and control strategies for helicopters can be found in [80] and [110].

The helicopter control team at the University of Technology of Compiègne, France, aims also at controlling the flight of a small helicopter in a three-dimensional space [37]. However, in order to gain insight into the modelling of the helicopter, the aerodynamic effects, the ground effect, the various interfaces and the sensors, we have built a vertical flying stand that allows the helicopter to move freely in the vertical axis as well as to turn around the vertical axis (i.e. the yaw). This means that the pitch and roll are mechanically constrained to be zero. The vertical platform is depicted in Figure 6.1.

A full model of a small helicopter mounted on this platform was obtained in [10] using a Lagrangian formulation and including the aerodynamic effects. This model has three DoF which are the altitude, the yaw and the rotor displacement. We have carried out a set of experiments to validate such a model. The resulting model is, in general, nonlinear since some of the parameters depend on the square of the main rotor velocity. It is clear that when the helicopter evolves in the flying stand, the main rotor velocity varies only slightly. Neglecting the rotor displacement leads to a system having two inputs (main rotor thrust and tail rotor thrust) and two outputs (altitude and yaw). It is natural to use the main rotor thrust to control the altitude and the tail rotor thrust to control the yaw. This suggested the use of a control algorithm based on pole-placement techniques [7, 91] for each one of the variables, i.e. altitude and yaw. Before computing the pole-placement controller, we compensate for the gravity term using the main rotor thrust, and compensate the coupling terms in the yaw dynamic equation using the tail rotor thrust.

We have carried out the synthesis of a pole-placement controller assuming the parameters are constant. However, when validating the model using experimental data, we have noticed some difficulties in estimating the values of the leading coefficients in the dynamic equations, i.e. those multiplying the control inputs. To take into account such parameter uncertainties, we have used a Lyapunov approach to show that the proposed controller is robust with respect to small uncertainties in the parameter values. Experiments have shown that the proposed non-adaptive controller remains stable, but presents a small oscillatory behaviour and the altitude fails to reach its desired value.

Controlling the altitude is important especially during take-off and landing of the helicopter. In order to improve the proposed altitude controller, we have developed an adaptive controller. The adaptation is only made on the leading coefficient corresponding to the main rotor thrust. We have used a Lyapunov analysis to prove that all the variables remain bounded and the altitude and yaw converge to their desired values.

We have noticed experimentally that it is very hard to make the transition from a non-adaptive controller to an adaptive one when the estimates move rapidly in a large domain. For security reasons we have to introduce some fixes in the adaptive controller to obtain a smooth transition between a non-adaptive scheme and the adaptive controller. We have constrained the parameters to belong to pre-specified intervals and we have introduced a gain into the gradient adaptation algorithm [77] to reduce, if required, the speed of adaptation of the parameters. The converge analysis has been carried out including these two modifications.

Experimental results showed that the adaptive controller improves the performance of the pole-placement algorithm. The adaptive controller is a direct type scheme using a gradient type adaptation algorithm. The obtained results seem useful for landing and taking-off of small helicopters.

This chapter is organized as follows: Section 6.2 describes the 2 DoF system which is used in Section 6.3 for the adaptive control design. The performance of the proposed adaptive controller is evaluated in real-time experiments in Section 6.4. Some concluding remarks end this chapter.

6.2 Dynamic Model

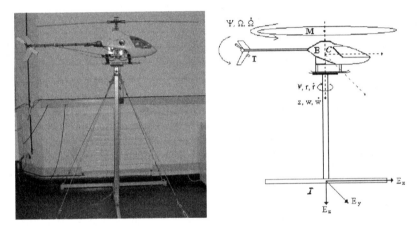

Fig. 6.1. Helicopter – vertical platform.

The model of a helicopter in a vertical flying stand (Figure 6.1) can be obtained using a Lagrangian formulation [10, 43]. Details on helicopter dynamics can be found in [78, 128, 141]. The vector of generalized coordinates (altitude, yaw and blade rotor azimuth) is given by $q = [z \quad \psi \quad \Psi]^T$, and $u = [\theta_0 \quad \theta_t]^T$ represents the vector of the control inputs: collective pitch and tail rotor collective respectively. The vertical force, the yaw torque, the main rotor torque and the rotor angular speed are denoted by f_z, τ_z, τ_ψ and Ω respectively. The c_i coefficients below (where $i = 1, ..., 12$), are related to physical ($i = 1, ..., 4$) and dynamic ($i = 5, ..., 12$) parameters of the helicopter [9, 43]. The dynamic model has the form

$$M(q)\ddot{q} + C(q, \dot{q})\dot{q} + G(q) = F(u) \qquad (6.1)$$

where M represents the inertia matrix

$$M(q) = \begin{bmatrix} c_0 & 0 & 0 \\ 0 & c_1 & c_2 \\ 0 & c_2 & c_3 \end{bmatrix} \tag{6.2}$$

C is the Coriolis matrix

$$C(q, \dot{q}) = \begin{bmatrix} c_{\omega M} & 0 & 0 \\ 0 & c_{\omega T}\Omega & 0 \\ 0 & 0 & 0 \end{bmatrix} \tag{6.3}$$

G denote the vector of conservative forces

$$G(q) = \begin{bmatrix} c_4 \\ 0 \\ 0 \end{bmatrix} \tag{6.4}$$

and F is the vector of generalized forces

$$F(u) = \begin{bmatrix} f_z \\ \tau_z \\ \tau_\Psi \end{bmatrix} = \begin{bmatrix} c_5\Omega^2\theta_0 + c_6\Omega + c_7 \\ c_8\Omega^2\theta_t \\ (c_9\Omega^2 + c_{10})\theta_0 + c_{11}\Omega^2 + c_{12} \end{bmatrix} \tag{6.5}$$

In the above equations the term $c_5\Omega^2\theta_0$ represents the main rotor thrust, $c_{\omega M}\omega$ corresponds to the main rotor inflow, $c_8\Omega^2\theta_t$ represents the tail rotor thrust, τ_Ψ contains the induced rotor torque and the profile rotor torque, $c_{\omega T}\Omega^2$ represents the tail rotor inflow. The various coefficients in (6.1)–(6.5) are specified in Appendix A. The helicopter dynamical equations (6.1)–(6.5) have been obtained by making several assumptions on the aerodynamic behaviour in order to simplify the model for control proposes. Equations (6.1)–(6.5) consider the rotor displacement as a DoF. However, we have experimentally observed that the small helicopter takes off at 1550 r.p.m. and that the rotor angular speed remains practically constant for the vertical displacement allowed in the platform. From model (6.1)–(6.5) the dynamics equation for the altitude can be rewritten as

$$m\dot{w} = c_5\Omega^2\theta_0 - mg \tag{6.6}$$

where m is the helicopter mass (i.e. 7.5 kg), $\Omega \approx 1550$ r.p.m. is the rotor angular speed, θ_0 is the blades collective pitch angle, g is the gravitational acceleration and the terms $c_6\Omega$, c_7 and $c_{\omega M}$ are neglected since they are relatively small.

From the second row of model (6.1)–(6.5), the yaw dynamics is given by

$$(c_1c_3 - c_2^2)\dot{r} = c_3c_8\Omega^2(\theta_t - k_g r) - c_2[(c_8\Omega^2 + c_{10})\theta_0 + c_{11}\Omega^2 + c_{12}] \tag{6.7}$$

where the term $-k_g r$ comes from the in-built angular velocity feedback of the gyro control system of the helicopter. This angular velocity feedback is part of the helicopter and has not been removed.

To simplify, equations (6.6) and (6.7) can be rewritten as

$$\dot{w} = k_1\theta_0 - k_2 \tag{6.8}$$
$$\dot{r} = k_3\theta_t - k_4 r - k_5\theta_0 - k_6 \tag{6.9}$$

where

$$k_1 = \frac{c_5\Omega^2}{m} \tag{6.10}$$

$$k_2 = g \tag{6.11}$$

$$k_3 = \frac{c_3 c_8 \Omega^2}{c_1 c_3 - c_2^2} \tag{6.12}$$

$$k_4 = \frac{c_3 c_8 \Omega^2 k_g}{c_1 c_3 - c_2^2} \tag{6.13}$$

$$k_5 = \frac{c_2(c_8\Omega^2 + c_{10})}{c_1 c_3 - c_2^2} \tag{6.14}$$

$$k_6 = \frac{c_2(c_{11}\Omega^2 + c_{12})}{c_1 c_3 - c_2^2} \tag{6.15}$$

The rotation motion of the yaw dynamics is controlled by a classical pole-placement method, it is computed in (6.9) in such a way that

$$\dot{r} = -k_7 r - k_8(\psi - \psi^d) \tag{6.16}$$

where k_7 and k_8 are positive constants and ψ^d is the desired yaw angle. Therefore, θ_t is given by

$$\theta_t = \frac{1}{k_3}[(k_4 - k_7)r + k_5\theta_0 + k_6 - k_8(\psi - \psi^d)] \tag{6.17}$$

The vertical position can also be controlled by using pole-placement. However, altitude control is critical especially when taking-off or landing the helicopter. We have noticed in practice that it is hard to estimate off-line the coefficient k_1 in equation (6.8). Notice that k_1 is related to c_5 (6.10). We can however estimate an upper bound and a lower bound for k_1. It is well known that an integrator control input can be used to cancel steady-state errors. Nevertheless, when performing real-time experiments with a helicopter, preventing accidents is the main concern. Introducing an integral in the control loop represents risks. Furthermore, the integral action does not converge when the desire altitude is time-varying. We therefore concluded that an adaptive controller should be used to stabilize the altitude of the helicopter.

6.3 Adaptive Altitude Robust Control Design

In this section, we propose an adaptive controller for the altitude of the helicopter. We will assume that the parameter k_1 in (6.8) belongs to the interval $[\bar{k}_1 - \Delta k_1, \bar{k}_1 + \Delta k_1]$, where $\bar{k}_1$ and Δk_1 are known. In other words, we will consider the second order system

$$\dot{w} = k_1 \theta_0 - k_2 \qquad (6.18)$$

where

$$k_1 = \bar{k}_1 + \Delta k_1 \qquad (6.19)$$

We use the pole-placement control law

$$\theta_0 = \alpha_1 w + \alpha_2 \left(z - z^d\right) + \hat{\theta} \qquad (6.20)$$

where α_1 and α_2 are positive constants, z^d is the desired altitude and $\hat{\theta}$ represents the estimate of

$$\theta = \frac{k_1}{k_2} \qquad (6.21)$$

Define the parameter error

$$\tilde{\theta} = \theta - \hat{\theta} \qquad (6.22)$$

then

$$\dot{\tilde{\theta}} = -\dot{\hat{\theta}} \qquad (6.23)$$

Introducing (6.19) and (6.20) into (6.18) we obtain

$$\dot{w} = \left(\bar{k}_1 + \Delta k_1\right) \alpha_1 w + \left(\bar{k}_1 + \Delta k_1\right) \alpha_2 \left(z - z^d\right) + k_1 \hat{\theta} - k_2 \qquad (6.24)$$

From (6.19) and (6.21) we get

$$\bar{\theta} - \Delta_\theta \le \theta \le \bar{\theta} + \Delta_\theta \qquad (6.25)$$

where

$$\bar{\theta} = \frac{\bar{k}_1}{k_2} \quad \text{and} \quad \Delta_\theta = \frac{\Delta k_1}{k_2} \qquad (6.26)$$

Define

$$a_1 = \alpha_1 \bar{k}_1 \qquad (6.27)$$
$$a_2 = \alpha_2 \bar{k}_1 \qquad (6.28)$$

Then, equation (6.24) can be rewritten as

$$\dot{w} = a_1 w + a_2 \left(z - z^d\right) + \Delta k_1 \left[\alpha_1 w + \alpha_2 \left(z - z^d\right)\right] + k_1 \left(\hat{\theta} - \frac{k_2}{k_1}\right) \quad (6.29)$$

Let us choose without loss of generality

$$a_1 = -1 \qquad\qquad (6.30)$$
$$a_2 = -1 \qquad\qquad (6.31)$$

Then, equation (6.29) becomes

$$\dot{w} + w + \left(z - z^d\right) = \Delta k_1 \left[\alpha_1 w + \alpha_2 \left(z - z^d\right)\right] - k_1 \tilde{\theta} \qquad (6.32)$$

where $\tilde{\theta}$ is given in (6.22). Define the following state variables

$$x_1 = z - z^d \qquad\qquad (6.33)$$
$$\dot{x}_1 = w \qquad\qquad (6.34)$$
$$x_2 = w \qquad\qquad (6.35)$$

Then, (6.32) can be rewritten as

$$\dot{x} = \mathbf{A}x + \mathbf{B}v \qquad\qquad (6.36)$$

where

$$x = \begin{bmatrix} x_1 \\ x_2 \end{bmatrix} \qquad\qquad (6.37)$$

$$\mathbf{A} = \begin{bmatrix} 0 & 1 \\ -1 & -1 \end{bmatrix} \qquad\qquad (6.38)$$

$$\mathbf{B} = \begin{bmatrix} 0 \\ 1 \end{bmatrix} \qquad\qquad (6.39)$$

$$v = \Delta k_1 \left[\alpha_1 w + \alpha_2 \left(z - z^d\right)\right] - k_1 \tilde{\theta} \qquad\qquad (6.40)$$

Consider the next candidate Lyapunov function

$$V = x^T \mathbf{P} x + \frac{1}{2} \frac{k_1 \tilde{\theta}^2}{\beta} \qquad\qquad (6.41)$$

where $\beta > 0$ is a coefficient used to regulate the parameter adaptation rate and $\mathbf{P} \in \mathbb{R}^{2 \times 2}$ is the following positive-definite matrix, which has been arbitrarily chosen as

$$\mathbf{P} = \begin{bmatrix} 3 & 1 \\ 1 & 2 \end{bmatrix} \qquad\qquad (6.42)$$

Using (6.36), the time derivative of V is given by

$$\dot{V} = \left(\mathbf{A}^T x^T + \mathbf{B}^T v\right)\mathbf{P}x + x^T\mathbf{P}(\mathbf{A}x + \mathbf{B}v) + \frac{k_1\tilde{\theta}\dot{\tilde{\theta}}}{\beta} \tag{6.43}$$

$$= x^T\left(\mathbf{A}^T\mathbf{P} + \mathbf{P}\mathbf{A}\right)x + \mathbf{B}^T v\mathbf{P}x + x^T\mathbf{P}\mathbf{B}v + \frac{k_1\tilde{\theta}\dot{\tilde{\theta}}}{\beta} \tag{6.44}$$

$$= x^T\left(\mathbf{A}^T\mathbf{P} + \mathbf{P}\mathbf{A}\right)x + 2x^T\mathbf{P}\mathbf{B}v + \frac{k_1\tilde{\theta}\dot{\tilde{\theta}}}{\beta} \tag{6.45}$$

where

$$\mathbf{A}^T\mathbf{P} + \mathbf{P}\mathbf{A} = \begin{bmatrix} -2 & 0 \\ 0 & -2 \end{bmatrix} \tag{6.46}$$

$$\mathbf{P}\mathbf{B} = \begin{bmatrix} 1 \\ 2 \end{bmatrix} \tag{6.47}$$

Then, equation (6.45) becomes (see also (6.23), (6.33)–(6.35) and (6.40))

$$\dot{V} = -2x_1^2 - 2x_2^2 + 2x_1v + 4x_2v + \frac{k_1\tilde{\theta}\dot{\tilde{\theta}}}{\beta} \tag{6.48}$$

$$= -2w^2 - 2\left(z - z^d\right)^2 + 2[(z - z^d) + 2w]\left[\Delta k_1\left[\alpha_1 w + \alpha_2\left(z - z^d\right)\right] - k_1\tilde{\theta}\right]$$
$$+ \frac{k_1\tilde{\theta}\dot{\tilde{\theta}}}{\beta} \tag{6.49}$$

$$= -2w^2 - 2\left(z - z^d\right)^2 + \Delta k_1\left(z - z^d\right)^2(2\alpha_2) + \Delta k_1 w^2(4\alpha_1)$$
$$+ \Delta k_1\left(z - z^d\right)w(2\alpha_1 + 4\alpha_2) - k_1\tilde{\theta}\left[2\left[(z - z^d) + 2w\right] + \frac{\dot{\tilde{\theta}}}{\beta}\right] \tag{6.50}$$

Using the fact that

$$a \cdot b \leq \frac{a^2 + b^2}{2} \tag{6.51}$$

equation (6.50) can be rewritten as

$$\dot{V} \leq -2w^2 - 2\left(z - z^d\right)^2 + \Delta k_1\left(z - z^d\right)^2(\alpha_1 + 4\alpha_2) + \Delta k_1 w^2(5\alpha_1 + 2\alpha_2)$$
$$- k_1\tilde{\theta}\left[2\left[(z - z^d) + 2w\right] + \frac{\dot{\tilde{\theta}}}{\beta}\right] \tag{6.52}$$

Consider the parameter estimation law

$$\dot{\tilde{\theta}} = -\beta\left(2\left(z - z^d\right) + 4w\right) \tag{6.53}$$

Now, equation (6.52) becomes

$$\dot{V} \le -2w^2 - 2\left(z - z^d\right)^2 + \Delta k_1 \left(z - z^d\right)^2 (\alpha_1 + 4\alpha_2)$$
$$+ \Delta k_1 w^2 (5\alpha_1 + 2\alpha_2) \tag{6.54}$$

In order to obtain a negative definite function, we assume that the parameter uncertainty Δk_1 satisfies the constraints

$$\Delta k_1 (\alpha_1 + 4\alpha_2) < 2 - \epsilon \tag{6.55}$$
$$\Delta k_1 (5\alpha_1 + 2\alpha_2) < 2 - \epsilon \tag{6.56}$$

Introducing (6.55) and (6.56) into (6.54), we get

$$\dot{V} \le -\epsilon w^2 - \epsilon \left(z - z^d\right)^2 \tag{6.57}$$

Using the La Salle theorem we conclude that

$$z \to z^d \quad \text{as} \quad t \to \infty \tag{6.58}$$
$$w \to 0 \quad \text{as} \quad t \to \infty \tag{6.59}$$

From (6.41) and (6.57) it follows that z, w and $\hat{\theta}$ are bounded. However, notice that $\hat{\theta}$ may not belong to the interval $[\bar{\theta} - \Delta_\theta, \bar{\theta} + \Delta_\theta]$ in which θ is known to lie. Experiments have shown that if we use $\hat{\theta}$ in (6.53) directly in the control law (6.20), it produces large oscillations in the altitude and the experiments have to be stopped before convergence of the altitude to the desired value. In order to avoid this problem we will project the estimate of $\hat{\theta}$ into the interval $[\bar{\theta} - \Delta_\theta, \bar{\theta} + \Delta_\theta]$. The projection procedure should be such that $\dot{V}$ in (6.57) still verifier $\dot{V} \le 0$. Let us consider again the following term in the RHS of (6.52):

$$k_1 \tilde{\theta} \left(y - \dot{\hat{\theta}}\right) \tag{6.60}$$

where

$$y = -\beta \left(2 \left(z - z^d\right) + 4w\right) \tag{6.61}$$

We know that

$$\bar{\theta} - \Delta_\theta \le \theta \le \bar{\theta} + \Delta_\theta \tag{6.62}$$

Indeed, we require the following two conditions to be verified:

C1: $\tilde{\theta} \left(y - \dot{\hat{\theta}}\right) \le 0$

C2: $\hat{\theta} \in \left[\bar{\theta} - \Delta_\theta, \bar{\theta} + \Delta_\theta\right]$

Using equation (6.22), condition C1 can be expressed as

$$\tilde{\theta}\left(y - \dot{\hat{\theta}}\right) = (\theta - \hat{\theta})\left(y - \dot{\hat{\theta}}\right) \tag{6.63}$$

$$= \left(\bar{\theta} + \delta - \hat{\theta}\right)\left(y - \dot{\hat{\theta}}\right) \leq 0 \tag{6.64}$$

where

$$|\delta| \leq \Delta_\theta \tag{6.65}$$

Both conditions C1 and C2 will be verified if $\hat{\theta}$ is chosen in the following way

$$\dot{\hat{\theta}} = \begin{cases} y & \text{for} \quad \hat{\theta} \in \left[\bar{\theta} - \Delta_\theta, \bar{\theta} + \Delta_\theta\right] \\ y & \text{for} \quad y > 0 \quad \text{and} \quad \hat{\theta} = \bar{\theta} - \Delta_\theta \\ 0 & \text{for} \quad y < 0 \quad \text{and} \quad \hat{\theta} = \bar{\theta} - \Delta_\theta \\ 0 & \text{for} \quad y > 0 \quad \text{and} \quad \hat{\theta} = \bar{\theta} + \Delta_\theta \\ y & \text{for} \quad y < 0 \quad \text{and} \quad \hat{\theta} = \bar{\theta} + \Delta_\theta \end{cases} \tag{6.66}$$

Therefore, we conclude that $z \rightarrow z^d$ when t goes towards infinity, and $\hat{\theta} \in [\bar{\theta} - \Delta_\theta, \bar{\theta} + \Delta_\theta]$.

6.4 Experimental Results

In this section, the performance of the proposed adaptive controller is examined and compared with that of a classical pole-placement controller.

6.4.1 Hardware

The radio-controlled helicopter used is a VARIO 1.8 m diameter rotor with a 23 cm^3 gasoline internal combustion engine. The radio is a Graupner MC-20. The vertical displacement is measured by a linear optical encoder and the yaw angle is obtained through a standard angular encoder. The radio and the PC (INTEL Pentium 3) are connected using data acquisition cards (ADVAN-TECH PCL-818HG and PCL-726). In order to simplify the experiments, the control inputs can be independently commuted between the automatic and the manual control modes. The connection in the radio is directly made to the joystick potentiometers for the gas and yaw controls. The vertical displacement of the helicopter in the platform varies from 1.8 to 2.5 m. Otherwise, it can turn freely around the vertical axis. The angular velocity feedback $-k_g r$ is carried out by the internal gyro control system of the helicopter.

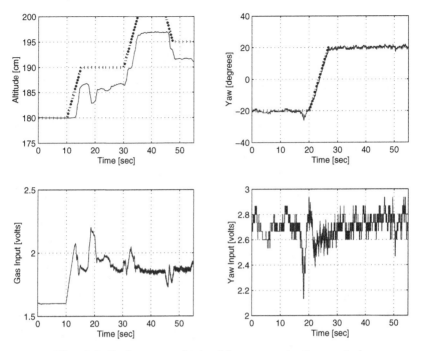

Fig. 6.2. Performance obtained for a non-adaptive control.

6.4.2 Experiment

We have computed estimates of the linear velocity w by using the approximation

$$w_t = \frac{z_t - z_{t-T}}{T} \quad (6.67)$$

where T is the sampling period, $T \approx \frac{1}{600}$. In order to obtain a good estimate of w and avoid abrupt changes in this signal, we have introduced a numerical filter. We have used a first order numerical filter. The gain and the pole of the filter were selected to improve the signal to noise ratio. The gain values used for the control law are $\beta = 0.01$, $\Delta_\theta = 0.22$, $\bar{\theta} = 1.6$, $a_1 = 0.01$ and $a_2 = 0.08$. The values of the gains in (6.8), (6.9) and (6.16) are $k_1 = 6.208$, $k_2 = 9.8$, $k_3 = 479.07$, $k_4 = 10$, $k_5 = 46.308$, $k_6 = 1269.76$, $k_7 = 0.1$ and $k_8 = 0.1$.

The experiment considers the case of the stabilization of the helicopter–platform dynamics for various values of the altitude and yaw. The desired altitude and yaw are given by the dotted lines in Figures 6.2 and 6.3. Figure 6.2 shows the behaviour of the altitude pole-placement controller when applied to the RC helicopter in the vertical flying stand for a non-adaptive controller given by

$$\theta_0 = 0.01 \times w - 0.08 \times \left(z - z^d\right) \tag{6.68}$$

We can see that the non-adaptive controller leads to oscillations and a considerable final error in the altitude dynamics. Figure 6.3 shows the performance of the adaptive altitude controller when applied to the RC helicopter. Note that we have no oscillations in the altitude dynamics and the final error is zero or very small.

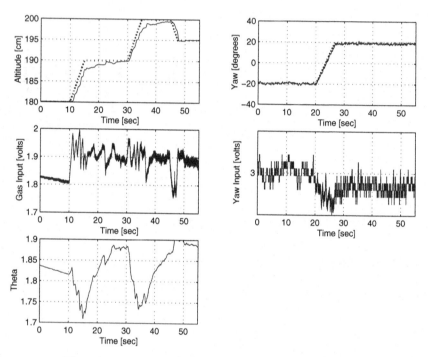

Fig. 6.3. Performance obtained for the adaptive control.

6.5 Conclusion

We have presented an adaptive control following an approach based on the adaptive pole-placement method. The algorithm controls the altitude dynamics of a real small-scale helicopter. The control algorithm has been proposed for a Lagrangian model of the RC helicopter. The proposed strategy has been successfully applied to a real helicopter in the flying stand.

7

Modelling and Control of a Tandem-Wing Tail-Sitter UAV

Dr. R. Hugh Stone,
University of Sydney, Australia

7.1 Introduction

Take-off and landing have historically presented difficulties for UAVs. If a runway is employed, much of the operational flexibility that is desired of a UAV is lost. Other solutions such as catapults or rocket assistance for take-off, and nets or parachutes for landing impose substantial costs and problems of their own. Although rotary wing UAVs are not subject to these landing and take-off problems they suffer performance limitations in terms of range, endurance and maximum forward speed. Other proposals aimed at combining some or all of the helicopter's low-speed flight characteristics with those of a normal aircraft include the tilt-rotor, tilt-wing and tilt-body. These vehicles, however, represent mechanically complex solutions, with attendant weight and cost penalties [159].

A potentially simpler solution for the UAV application (where there are no passengers and crew who like to stay upright) is the tail-sitter. In keeping with the basic simplicity of the tail-sitter configuration, hover control can be effected via normal wing-mounted control surfaces. Initial work on the T-Wing configuration envisaged transitioning from vertical to horizontal flight via a stall tumble manoeuvre, however recent work has shown that for typical vehicle configurations, an unstalled transition is also easily possible with only modest amounts of excess thrust. Furthermore, this particular flight profile dispenses with any requirement for large edge-on flows into the propeller disc and hence allows the use of normal variable-pitch propellers in preference to more complicated, fully articulated helicopter-like rotors.

7.2 Tail-Sitters: A Historical Perspective

Although tail-sitter vehicles have been investigated over the last 50 years as a means to combine the operational advantages of vertical flight enjoyed by helicopters with the better horizontal flight attributes of conventional airplanes, no successful tail-sitter vehicles have ever been produced. One of the primary reasons for this is that tail-sitters such as the Convair XF-Y1 and Lockheed XF-V1 (see Figure 7.1) experimental vehicles of the 1950s proved to be very difficult to pilot during vertical flight and the transition manoeuvres.

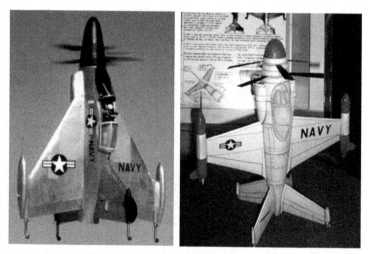

Fig. 7.1. Convair XF-Y1 and Lockheed XF-V1 Tail-Sitter Aircraft [22].

With the advent of modern computing technology and improvements in sensor reliability, capability and cost it is now possible to overcome these piloting disadvantages by transitioning the concept to that of an unmanned vehicle. With the pilot replaced by modern control systems it should be possible to realize the original promise of the tail-sitter configuration.

The tail-sitter aircraft considered in this chapter differs substantially from its earlier counterparts and is most similar in configuration to the Boeing Heliwing vehicle of the early 1990s. This vehicle had a 1450 lb maximum take-off weight (MTOW) with a 200 lb payload, 5 hour endurance and 180 kts maximum speed and used twin rotors powered by a single 240 SHP turbine engine [113]. A picture of the Heliwing is shown in Figure 7.2.

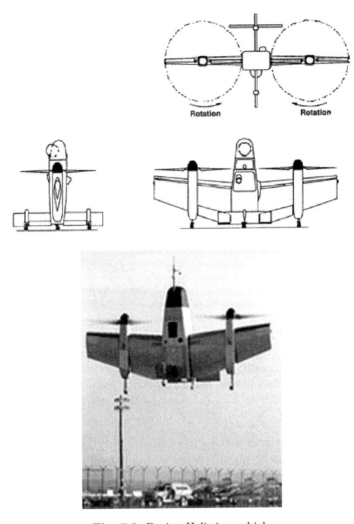

Fig. 7.2. Boeing Heliwing vehicle.

7.3 Applications for a Tail-Sitter UAV

Although conflicts over the last 20 years have demonstrated the importance of military UAV systems in facilitating real-time intelligence gathering, it is clear that most current systems still do not possess the operational flexibility that is desired by force commanders. One of the reasons for this is that most UAVs have adopted relatively conventional aircraft configurations. This leads directly to operational limitations because it either necessitates take-off and landing from large fixed runways, or the use of specialized launch and recovery methods such catapults, rockets, nets, parachutes and airbags.

One potential solution to these operational difficulties is a tail-sitter VTOL UAV. Such a vehicle has few operational requirements other than a small clear area for take-off and landing. While other VTOL concepts share this operational advantage over conventional vehicles, the tail-sitter has some other unique benefits. In comparison with helicopters, a tail-sitter vehicle does not suffer the same performance penalties in terms of dash-speed, range and endurance because it spends the majority of its mission in a more efficient airplane flight mode. The only other VTOL concepts that combine vertical and horizontal flight are the tilt-rotor and tilt-wing, however, both involve significant extra mechanical complexity in comparison with the tail-sitter vehicle, which has fixed wings and nacelles. A further simplification can be made in comparison with other VTOL designs by the use of prop-wash over wing and fin mounted control surfaces to effect control during vertical flight, thus obviating the need for cyclic rotor control [161].

7.3.1 Defence Applications

For naval forces, a tail-sitter VTOL UAV has enormous potential as an aircraft that can be deployed from small ships and used for long-range reconnaissance and surveillance; over-the-horizon detection of low-flying missiles and aircraft; deployment of remote acoustic sensors; and as a platform for aerial support and communications. The vehicle could also be used in anti-submarine activities and anti-surface operations and is ideal for battlefield monitoring over both sea and land. The obvious benefit in comparison with a conventional UAV is the operational flexibility provided by the vertical launch and recovery of the vehicle [162].

For ground-based forces a tail-sitter vehicle is also attractive because it allows UAV systems to be quickly deployed from small cleared areas with a minimum of support equipment. This makes the UAVs less vulnerable to attacks on fixed bases without the need to set-up catapult launchers or recovery nets. It is envisaged that ground forces would mainly use small VTOL UAVs as reconnaissance and communication relay platforms.

7.3.2 Civilian Applications

Besides the defence requirements, there are also many civilian applications for which a VTOL UAV is admirably suited. Coastal surveillance to protect national borders from illegal immigrants and illicit drugs is clearly an area where such vehicles could be used. The VTOL characteristics in this role are an advantage, as they allow such vehicles to be based in remote areas without the fixed infrastructure of airstrips, or to be operated from small coastal patrol vessels.

Further applications are also to be found in mineral exploration and environmental monitoring in remote locations. While conventional vehicles could of course accomplish such tasks, their effectiveness may be limited if forced to operate from bases a long way from the area of interest.

7.4 The T-Wing: A Tandem-Wing Tail-Sitter UAV

The T-Wing[1] vehicle is somewhat similar to the Boeing Heliwing of the early 1990s in that it also has twin wing-mounted propellers, however it differs from that vehicle in a number of important respects:

- In keeping with the basic simplicity of the tail-sitter configuration, control is effected via prop-wash over the wing and fin mounted control surfaces, rather than using helicopter cyclic control. This vehicle is similar to the early tail-sitter vehicles of the 1950s, the Convair XF-Y1 and the Lockheed XF-V1 (see Figure 7.1). Collective blade pitch control is still required to marry efficient high-speed horizontal flight performance with the production of adequate thrust on take-off, however, even this complication can be deleted, with little performance penalty, if high dash speeds are not required.

- The current vehicle uses a canard to allow a more advantageous placement of the vehicle centre of gravity (CG).

- Two separate engines are used in the current design though the possibility of using a single engine with appropriate drive trains could also be accommodated.

The basic configuration of the T-Wing is presented in Figure 7.3 and Figure 7.4, with a diagram in Figure 7.5 showing some of the important gross geometric properties. The aircraft is essentially a tandem wing configuration with twin tractor propellers mounted on the aft main wing.

The current T-Wing vehicle is a technology demonstrator and not a prototype production one. The aims of the T-Wing vehicle programme are to prove the critical technologies required of a tail-sitter vehicle before committing funds to full-scale development. The most important aspects of the T-Wing design that have to be demonstrated are reliable autonomous hover control and the ability to perform the transition manoeuvres between horizontal and vertical flight.

---

[1] This work is part of an ongoing research programme involving the construction and test of a concept demonstrator tail-sitter UAV. The *T-Wing* was built by the University of Sydney in collaboration with Sonacom Pty Ltd [158, 160].

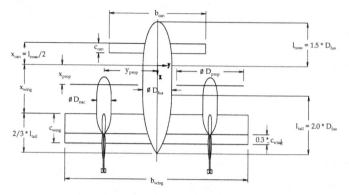

Fig. 7.3. Plan view of the T-Wing.

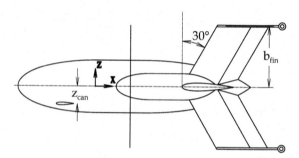

Fig. 7.4. Side view of the T-Wing.

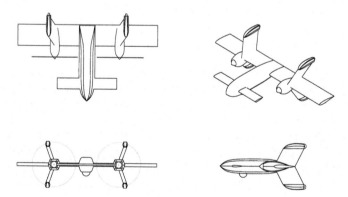

Fig. 7.5. T-Wing vehicle configuration.

7.5 Description of the T-Wing Vehicle

The T-Wing technology demonstrator vehicle has a nominal a 55 lb (25 kg) Maximum Take-off Weight (MTOW) with a 7 ft (2.13 m) wing span and a total length (from nose to fin tip) of 5 ft (1.52 m) [163, 164, 165, 166, 167].

The vehicle was originally designed to be powered by two geared 4.5 HP electric brushless DC motors driving 30 inch fixed pitch counter-rotating propellers and supplied by up to 20 lb (9.1 kg) of Ni-Cd batteries. This was designed to give the vehicle a maximum endurance of 5–6 minutes, which was long enough to accomplish the critical flight control objectives of demonstrating stable autonomous hover along with the two transition manoeuvres. The reason for initially selecting electric rather than petrol propulsion was because electric motors promised easier set-up and operation in comparison with petrol engines. Unfortunately problems with the particular electric motor speed controllers selected caused excessive delays and doubts about system reliability. For these reasons, it was decided in August 2000 to convert the vehicle to run on petrol engines [161].

The petrol-engined version uses two 6 HP 2-stroke motors and has the same nominal weight as the electric vehicle. The petrol engines drive two counter-rotating 23 inch fixed pitch propellers directly. For the sake of system simplicity there is no cross-shafting between the two engines. Due to the higher installed power of the petrol engines this vehicle has considerably more excess thrust than the electric vehicle and it is anticipated that the MTOW can be pushed to at least 65 lb (29.5 kg). Although the petrol vehicle is still very much a concept demonstration platform, this increased take-off weight should allow an endurance of up to several hours carrying a 5 lb payload.

The vehicle is built primarily of carbon-fibre and glass-fibre composite materials with local panel stiffness provided by the use of Nomex honeycomb core material. The airframe has been statically tested to a normal load factor in excess of 8 G's [73]. A picture of the completed T-Wing vehicle is shown in Figure 7.6.

During hover, the vehicle is controlled in *pitch* and *roll* via elevon control surfaces on the wing which are submerged in the prop-wash of the propellers. Yaw control of the vehicle is effected via fins and rudders attached to the nacelles and which are also submerged in the propeller slipstream. Additionally the tips of the fins provide the attachment point for the landing gear and hence determine the *footprint* of the vehicle on the ground.

7.5.1 Typical Flight Path for the T-Wing Vehicle

From the beginning of the T-Wing concept in mid-1995 it was proposed that the vehicle be allowed to transition from vertical to horizontal flight via a

Fig. 7.6. T-Wing demonstrator vehicle.

stall-tumble manoeuvre. This was seen as offering the potential advantage of allowing the vehicle to carry less excess hover thrust capability than would be required if a smooth installed transition was mandated. More recent studies [168] have however shown that unstalled vertical to horizontal transitions are also possible with only modest excess thrust levels ($< 20\%$). Because of this and the uncertainties associated with post-stall flight, current plans call for the use of the unstalled manoeuvre. The reverse transition manoeuvre is less novel and simply involves the vehicle performing a pull-up to regain a vertical attitude followed by a slow descent. The mission phase of flight is performed in the more efficient horizontal *aeroplane* mode of flight. This flight regime is shown in Figure 7.7.

The transition manoeuvres described above will affect the operational utility of the vehicle in a number of ways. Some of these are listed below.

- The transitions should take as little time to complete as possible to enable the vehicle to quickly enter and exit the mission phase of its flight regime.

- The vehicle should gain as little excess height as possible during these manoeuvres to minimize possible conflict with overlying airspace and in the military case to minimize the chance of being detected.

- The vehicle should if possible avoid excessive angle of attack regions where the simulation and analysis is uncertain.

- A last reason to also avoid stalled flight conditions or *tumble* manoeuvres is because they worry potential users.

With these considerations in mind some work has been done in performing numerical optimizations of the transition manoeuvres to try and achieve the above goals.

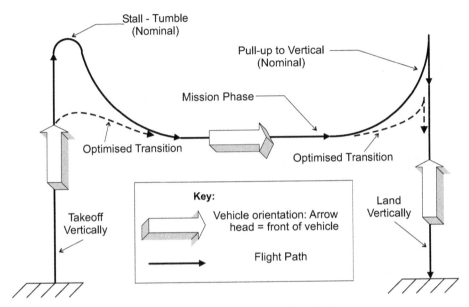

Fig. 7.7. T-Wing vehicle flight path.

7.6 6-DOF Nonlinear Model

The starting point for the work done on transition manoeuvre optimization as well as vehicle control design is a full nonlinear 6-DOF model of the T-Wing vehicle that has been developed by [160]. Basically this model consists of the normal nonlinear, rigid-body, 6-DOF equations of motion that apply to any aircraft (Figure 7.8); a simple mass model of the vehicle; and a large database of basic forces (or coefficients) and aerodynamic derivatives covering a large number of flight conditions.

The next subsection shows the derivation of the nonlinear model of an airplane, after which the equations are particularized to the case of the T-Wing.

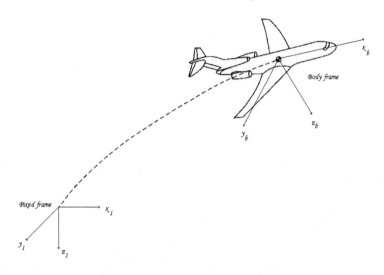

Fig. 7.8. Body and inertial axis system.

7.6.1 Derivation of Rigid Body Equations of Motion

The rigid body equations of motion are obtained from Newton's second law [21]:

$$\sum \mathbf{F} = \frac{d}{dt}(mv) \tag{7.1}$$

$$\sum \mathbf{M} = \frac{d}{dt}\mathbf{H} \tag{7.2}$$

- The summation of all external forces acting on a body is equal to the time rate of change of the momentum of the body.

- The summation of the external moments acting on the body is equal to the time rate of change of the moment of momentum (angular momentum).

  The vector equations above can be rewritten in scalar form as

$$F_x = \frac{d}{dt}(mu)$$
$$F_y = \frac{d}{dt}(mv) \tag{7.3}$$
$$F_z = \frac{d}{dt}(mw)$$

where F_x, F_y, F_z and u, v, w are the components of the force and velocity along the x, y and z axes respectively, and

$$L = \frac{d}{dt}H_x$$

$$M = \frac{d}{dt}H_y \qquad (7.4)$$

$$N = \frac{d}{dt}H_z$$

where L, M, N and H_x, H_y, H_z are the components of the moment and moment of momentum along the x, y and z axes, respectively.

The force components are composed of contributions due to the aerodynamic, propulsive and gravitational forces acting on the airplane.

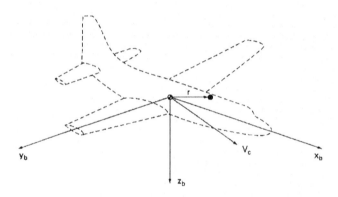

Fig. 7.9. An element of mass on an airplane.

If we take an element of mass δm of the airplane (Figure 7.9), then Newton's second law can be rewritten as

$$\delta \mathbf{F} = \delta m \frac{d\mathbf{v}}{dt} \qquad (7.5)$$

where $\mathbf{v}$ is the velocity of the elemental mass relative to an absolute or inertial frame, and $\delta \mathbf{F}$ is the resulting force acting on the elemental mass.

The total external force acting on the aircraft is found by summing all the elements of the aircraft:

$$\sum \delta \mathbf{F} = \mathbf{F} \qquad (7.6)$$

The velocity of the differential mass δm is

$$\mathbf{v} = \mathbf{v}_c + \frac{d\mathbf{r}}{dt} \tag{7.7}$$

where $\mathbf{v}_c$ is the velocity of the centre of mass of the airplane and $\frac{d\mathbf{r}}{dt}$ is the velocity of the element relative to the centre of mass.

Rewriting Newton's second law gives

$$\sum \delta \mathbf{F} = \frac{d}{dt} \sum \left(\mathbf{v}_c + \frac{d\mathbf{r}}{dt} \right) \delta m \tag{7.8}$$

If we assume that the mass of the vehicle is constant, equation (7.8) can be rewritten as

$$\mathbf{F} = m \frac{d\mathbf{v}_c}{dt} + \frac{d^2}{dt^2} \sum \mathbf{r}\, \delta m \tag{7.9}$$

In the above equations $\mathbf{r}$ is measured from the centre of mass, therefore the summation $\sum \mathbf{r}\, \delta m$ is equal to 0. Then, the external force on the airplane of the motion of the vehicle's centre of mass can be written as

$$\mathbf{F} = m \frac{d\mathbf{v}_c}{dt} \tag{7.10}$$

For the differential element of mass, δm, the moment equation can be written as

$$\delta \mathbf{M} = \frac{d}{dt} \delta \mathbf{H} = \frac{d}{dt}(\mathbf{r} \times \mathbf{v}) \delta m \tag{7.11}$$

The velocity of the mass element can be expressed as

$$\mathbf{v} = \mathbf{v}_c + \frac{d\mathbf{r}}{dt} = \mathbf{v}_c + \omega \times \mathbf{r} \tag{7.12}$$

where ω is the angular velocity of the vehicle and $\mathbf{r}$ is the position of the mass element measured from the centre of mass. We can express ω and $\mathbf{r}$ as

$$\omega = p\mathbf{i} + q\mathbf{j} + r\mathbf{k}$$
$$\mathbf{r} = x\mathbf{i} + y\mathbf{j} + z\mathbf{k}$$

The total moment of momentum can be written as

$$\mathbf{H} = \sum \delta \mathbf{H} = \sum \mathbf{r}\, \delta m \times \mathbf{v}_c + \sum [\mathbf{r} \times (\omega \times \mathbf{r})] \delta m \tag{7.13}$$

As explained previously, the term $\sum \mathbf{r}\, \delta m$ is 0. Rewriting $\mathbf{H}$ in its scalar components we have

$$H_x = p \sum (y^2 + z^2)\delta m - q \sum xy\,\delta m - r \sum xz\,\delta m$$

$$H_y = -p \sum xy\,\delta m + q \sum (x^2 + z^2)\,\delta m - r \sum yz\,\delta m \qquad (7.14)$$

$$H_z = -p \sum xz\,\delta m - q \sum yz\,\delta m + r \sum (x^2 + y^2)\,\delta m$$

The summations in these equations are the mass moment and products of inertia of the airplane. Rewriting the above equations in terms of I_{xx}, I_{yy} and I_{zz} (mass moments of inertia of the body about the x, y and z axes) and in terms of the products of inertia (the terms with mixed indices) gives

$$H_x = pI_x - qI_{xy} - rI_{xz}$$

$$H_y = -pI_{xy} + qI_y - rI_{yz} \qquad (7.15)$$

$$H_z = -pI_{xz} - qI_{yz} + rI_z$$

If the reference frame is not rotating, then as the airplane rotates the moments and products of inertia will vary with time. To avoid this difficulty we will fix the axis system to the aircraft (body axis system).

Taking the derivatives of the vectors $\mathbf{v}$ and $\mathbf{H}$ referred to the rotating body frame of reference, we have

$$\mathbf{F} = m\frac{d\mathbf{v}_c}{dt}\bigg|_B + m(\omega \times \mathbf{v}_c) \qquad (7.16)$$

$$\mathbf{M} = \frac{d\mathbf{H}}{dt}\bigg|_B + \omega \times \mathbf{H} \qquad (7.17)$$

where the subscript B refers to the body fixed frame of reference.

Rewriting the above vector equations in scalar form, we obtain

$$F_x = m(\dot{u} + qw - rv)$$

$$F_y = m(\dot{v} + ru - pw)$$

$$F_z = m(\dot{w} + pv - qu)$$

$$L = \dot{H}_x + qH_z - rH_y$$

$$M = \dot{H}_y + rH_x - pH_z$$

$$N = \dot{H}_z + pH_y - qH_x$$

The components of the force and moment acting on the airplane are composed of aerodynamic, gravitational, and propulsive contributions.

By positioning the body axis xz plane in the (usual) aircraft plane of symmetry, the xy and yz products of inertia can be rewritten as $I_{yz} = I_{xy} = 0$. With this assumption, the moment equations can be written as

$$L = I_x \dot{p} - I_{xz}\dot{r} + qr(I_z - I_y) - I_{xz}pq$$
$$M = I_y \dot{q} + rp(I_x - I_z) + I_{xz}(p^2 - r^2) \qquad (7.18)$$
$$N = -I_{xz}\dot{p} + I_z\dot{r} + pq(I_y - I_x) + I_{xz}qr$$

These equations represent the general equations for airplanes. For the T-Wing aircraft, which also possesses a plane of symmetry in the xy plane, the last product of inertia I_{xz} is 0.

7.6.2 Orientation of the Aircraft

The equations of motion have been derived for an axis system fixed to the airplane. Unfortunately, the position and orientation of the airplane cannot be described relative to the moving body axis frame. Instead, the orientation and position of the airplane must be referenced to a fixed frame.

The orientation of typical aircraft can be described by three consecutive rotations, whose order is important. The angular rotations are called the Euler angles. They consist of the following ordered rotations:

- Start from a horizontal attitude in which the vehicle x_b (forward) axis points North and the vehicle z_b (belly) axis points down. In this starting position the vehicle axes are aligned with the standard North-East-Down (NED) axes.

- Apply a yaw rotation (ψ) about the z_b axis.

- Apply a pitch rotation (θ) about the new position of the y_b axis.

- Finally perform a roll rotation (ϕ) about the twice moved x_b axis.

A complicating factor in the dynamic analysis of this vehicle, compared with other aircraft, is the fact that it spends a considerable time in a predominantly vertical attitude. While in this state the use of the normal Euler angle attitude representation is inappropriate as the yaw (ψ) and roll (ϕ) angles are not uniquely determined when the pitch angle (θ) is 90°. As well, the Euler angles have *wrap-around* difficulties when the yaw and roll angles exceed their bounds of ±180°. To fix these deficiencies a second *vertical* set of Euler angles are defined. This set of Euler angles is constructed to provide unambiguous attitudes in the vertical flight orientation. These vertical Euler angles and the order of their application are defined as follows:

- Start from a vertical attitude in which the vehicle x_b axis points upwards (i.e. opposite to the earth fixed z_e axis in the standard North-East-Down (NED) system) and with the z_b axis pointing North.

- Apply a roll rotation (ϕ_v) about the x_b axis.

- Apply a pitch rotation (θ_v) about the new position of the y_b axis.

- Finally perform a yaw rotation (ψ_v) about the twice moved z_b axis.

Note that in this system the order of rotations is roll, pitch, yaw (ϕ_v, θ_v, ψ_v) rather than yaw, pitch, roll (ϕ, θ, ψ) as in the standard Euler angle rotations. It should also be noted that this system gives singularities when the vertical pitch angle, θ_v, is $\pm 90°$: in other words when the vehicle is horizontal. These vertical Euler angles (Figure 7.10) can be used in conjunction with the normal Euler angles and quaternions to give smooth attitude representations throughout the full-flight envelope while still allowing physical insight and the use of feedback control at any attitude. The three distinct systems of attitude representation are summarized below.

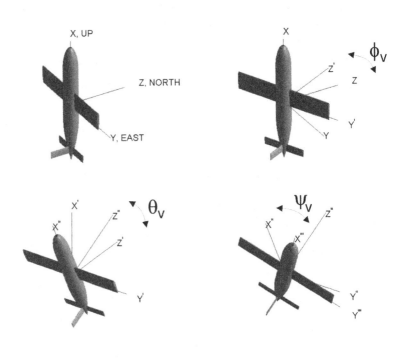

Fig. 7.10. Definition of vertical Euler angles.

QUATERNION REPRESENTATION

The four quaternion parameters $(q_0,\ q_1,\ q_2,\ q_3)$ are used internally in the 6-DOF simulation to avoid the possible computational errors associated with either of the Euler angle representations.

NORMAL EULER ANGLES

The normal Euler angles $(\psi,\ \theta,\ \phi)$ are used as *output and feedback* parameters while the vehicle is in a substantially horizontal attitude: typically while the vehicle pitch angle is less than 45° $(\theta \leq 45°;\ \theta_v \geq 45°)$.

VERTICAL EULER ANGLES

The vertical Euler angles $(\phi_v,\ \theta_v,\ \psi_v)$ are used as output and feedback parameters while the vehicle is in a substantially vertical attitude: typically while the vehicle pitch angle is greater than 45° $(\theta > 45°;\ \theta_v < 45°)$.

Having defined the Euler angles, one can determine the flight velocity components relative to the fixed reference frame.

Transformations between attitude representations

In the following transformations the earth-fixed axes are a North-East-Down (NED) system. In this system the x_e axis points North, the y_e axis points east, and the z_e axis points down to the centre of the earth.

TRANSFORMATIONS FROM NED FRAME TO VEHICLE BODY AXIS FRAME

In the following equations $\mathbf{B} = \{b_{ij}\}$ is the transformation matrix that takes vectors from the NED frame to the body axis frame. As $\mathbf{B}$ is orthogonal, its inverse is its transpose, $\mathbf{B}^{-1} = \mathbf{B}^T$.

Before proceeding further, let us use the shorthand notation $S_\psi \equiv \sin(\psi)$, $C_\psi \equiv \cos(\psi)$, $S_\theta \equiv \sin(\theta)$, and so forth.

USING NORMAL EULER ANGLES

$$\mathbf{B} = \begin{bmatrix} C_\theta C_\psi & C_\theta S_\psi & -S_\theta \\ -C_\phi S_\psi + S_\phi S_\theta C_\psi & C_\phi C_\psi + S_\phi S_\theta S_\psi & S_\phi C_\theta \\ S_\phi S_\psi + C_\phi S_\theta C_\psi & -S_\phi C_\psi + C_\phi S_\theta S_\psi & C_\phi C_\theta \end{bmatrix} \qquad (7.19)$$

Using vertical Euler angles

$$
\mathbf{B} =
\begin{bmatrix}
-C_{\psi_v}S_{\theta_v}C_{\phi_v} + S_{\psi_v}S_{\phi_v} & C_{\psi_v}S_{\theta_v}S_{\phi_v} + S_{\psi_v}C_{\phi_v} & -C_{\psi_v}C_{\theta_v} \\[2mm]
S_{\psi_v}S_{\theta_v}C_{\phi_v} + C_{\psi_v}S_{\phi_v} & -S_{\psi_v}S_{\theta_v}S_{\phi_v} + C_{\psi_v}C_{\phi_v} & S_{\psi_v}C_{\theta_v} \\[2mm]
C_{\theta_v}C_{\phi_v} & -C_{\theta_v}S_{\phi_v} & -S_{\theta_v}
\end{bmatrix}
\tag{7.20}
$$

Using quaternions

$$
\mathbf{B} =
\begin{bmatrix}
q_0^2 + q_1^2 - q_2^2 - q_3^2 & 2(q_1q_2 + q_0q_3) & 2(q_1q_3 - q_0q_2) \\[2mm]
2(q_1q_2 - q_0q_3) & q_0^2 - q_1^2 + q_2^2 - q_3^2 & 2(q_2q_3 + q_0q_1) \\[2mm]
2(q_1q_3 + q_0q_2) & 2(q_2q_3 - q_0q_1) & q_0^2 - q_1^2 - q_2^2 + q_3^2
\end{bmatrix}
\tag{7.21}
$$

Quaternions in terms of normal Euler angles

$$
\begin{aligned}
q_0 &= \cos(\tfrac{\phi}{2})\cos(\tfrac{\theta}{2})\cos(\tfrac{\psi}{2}) + \sin(\tfrac{\phi}{2})\sin(\tfrac{\theta}{2})\sin(\tfrac{\psi}{2}) \\[2mm]
q_1 &= \sin(\tfrac{\phi}{2})\cos(\tfrac{\theta}{2})\cos(\tfrac{\psi}{2}) - \cos(\tfrac{\phi}{2})\sin(\tfrac{\theta}{2})\sin(\tfrac{\psi}{2}) \\[2mm]
q_2 &= \cos(\tfrac{\phi}{2})\sin(\tfrac{\theta}{2})\cos(\tfrac{\psi}{2}) + \sin(\tfrac{\phi}{2})\cos(\tfrac{\theta}{2})\sin(\tfrac{\psi}{2}) \\[2mm]
q_3 &= \cos(\tfrac{\phi}{2})\cos(\tfrac{\theta}{2})\sin(\tfrac{\psi}{2}) - \sin(\tfrac{\phi}{2})\sin(\tfrac{\theta}{2})\cos(\tfrac{\psi}{2})
\end{aligned}
\tag{7.22}
$$

Quaternions in terms of vertical Euler angles

$$
\begin{aligned}
q_0 &= \sqrt{\tfrac{1-\sin(\theta_v)}{2}} \, \cos\left(\tfrac{\psi_v - \phi_v}{2}\right) \\[2mm]
q_1 &= \sqrt{\tfrac{1+\sin(\theta_v)}{2}} \, \sin\left(\tfrac{\psi_v + \phi_v}{2}\right) \\[2mm]
q_2 &= \sqrt{\tfrac{1+\sin(\theta_v)}{2}} \, \cos\left(\tfrac{\psi_v + \phi_v}{2}\right) \\[2mm]
q_3 &= \sqrt{\tfrac{1-\sin(\theta_v)}{2}} \, \sin\left(\tfrac{\psi_v - \phi_v}{2}\right)
\end{aligned}
\tag{7.23}
$$

Vertical Euler angles in terms of normal Euler angles

$$\theta_v = -\sin^{-1}(\cos(\phi)\ \cos(\theta)) \hspace{3cm} = \sin^{-1}(-b_{33})$$

$$\psi_v = \sin^{-1}\left(\frac{\sin(\phi)\ \cos(\theta)}{\cos(\theta_v)}\right) \hspace{2cm} = -\tan^{-1}\left(\frac{b_{23}}{b_{13}}\right)$$

$$\phi_v = -\sin^{-1}\left(\frac{-\sin(\phi)\ \cos(\psi)+\ \cos(\phi)\ \sin(\theta)\ \sin(\psi)}{\cos(\theta_v)}\right) = -\tan^{-1}\left(\frac{b_{32}}{b_{31}}\right)$$

$$(7.24)$$

Normal Euler angles in terms of vertical Euler angles

$$\theta = \sin^{-1}(\cos(\psi_v)\ \cos(\theta_v)) \hspace{3cm} = -\sin^{-1}(b_{13})$$

$$\phi = \sin^{-1}\left(\frac{\sin(\psi_v)\ \cos(\theta_v)}{\cos(\theta)}\right) \hspace{2cm} = \tan^{-1}\left(\frac{b_{23}}{b_{33}}\right)$$

$$\psi = \sin^{-1}\left(\frac{\cos(\psi_v)\ \sin(\theta_v)\ \sin(\phi_v)+\ \sin(\psi_v)\ \cos(\phi_v)}{\cos(\theta)}\right) = \tan^{-1}\left(\frac{b_{12}}{b_{11}}\right)$$

$$(7.25)$$

Gravitational and thrust forces

The gravitational force acting on the aircraft acts through the centre of gravity of the aircraft. Because the body axis system is fixed to the centre of gravity, the gravitational force will not produce any moments. It will contribute to the external force acting on the aircraft, however, and have components along the respective body axes.

In a general case, the thrust force due to the propulsion system can have components that act along each of the body axis directions. In addition, the propulsive forces also can create moments if the thrust does not act through the centre of gravity. In the case of the present T-Wing vehicle however, the thrust axis is aligned with the aircraft x_b axis and coincident with the centre of gravity.

Finally, we present a summary of the rigid body equations of motion.

7.6.3 Equations of Motion

The general 6-DOF equations of motion for a flight vehicle in a flat-earth body axis system are given by the following equations:

FORCE EQUATIONS

$$\dot{u} = rv - qw - g_x + \frac{F_x}{m} \tag{7.26}$$

$$\dot{v} = -ru + pw + g_y + \frac{F_y}{m} \tag{7.27}$$

$$\dot{w} = qu - pv - g_z + \frac{F_z}{m} \tag{7.28}$$

MOMENTS EQUATIONS

$$
\begin{aligned}
L &= I_x\dot{p} - I_{xz}\dot{r} + qr(I_z - I_y) - I_{xz}pq \\
M &= I_y\dot{q} + rp(I_x - I_z) + I_{xz}(p^2 - r^2) \\
N &= -I_{xz}\dot{p} + I_z\dot{r} + pq(I_y - I_x) + I_{xz}qr
\end{aligned}
\tag{7.29}
$$

KINEMATIC EQUATIONS

$$
\begin{bmatrix} \dot{q}_0 \\ \dot{q}_1 \\ \dot{q}_2 \\ \dot{q}_3 \end{bmatrix}
= -\frac{1}{2}
\begin{bmatrix}
0 & p & q & r \\
-p & 0 & -r & q \\
-q & r & 0 & -p \\
-r & -q & p & 0
\end{bmatrix}
\begin{bmatrix} q_0 \\ q_1 \\ q_2 \\ q_3 \end{bmatrix}
\tag{7.30}
$$

NAVIGATION EQUATIONS

$$
\begin{bmatrix} \dot{p}_N \\ \dot{p}_E \\ -\dot{h} \end{bmatrix}
= \mathbf{B}^{-1}
\begin{bmatrix} u \\ v \\ w \end{bmatrix}
\tag{7.31}
$$

where u, v, w are body axis velocities in the x, y and z directions respectively, while p, q, r are the body axis rates about these same axes. The moments of inertia are I_{xx}, I_{yy}, I_{zz}, I_{xz} and in the case of the T-Wing, I_{xz} is zero, which produces further simplifications. F_x, F_y and F_z are the forces in the body axis directions, while L, M and N are the moments about these axes (including any gyroscopic components[2]). p_N, p_E and h are the North, East and height

---

[2] As the propellers on the T-Wing are contra-rotating, gyroscopic moments will essentially be zero. The only case in which gyrsocopic moments will not be zero is if there is a significant difference in the RPM of the two engines. This may occur, for instance, if the vehicle seeks to maintain trimmed hover flight in the presence of a strong sideways cross-wind via the use of differential throttles. Even in this case the difference in the engine RPMs will typically be relatively minor and hence the gyroscopic moments small.

coordinates in the fixed earth frame. $\mathbf{B}$ is the transformation matrix from the NED frame to the body axis frame given above (in any representation), and g_x, g_y and g_z are the gravity components in the body axis frame (see equation (7.32)), while g_0 is the local acceleration due to gravity.

$$
\begin{bmatrix} g_x \\ g_y \\ g_z \end{bmatrix} = \mathbf{B} \begin{bmatrix} 0 \\ 0 \\ g_0 \end{bmatrix} \tag{7.32}
$$

In this last equation g_0 is the acceleration due to gravity. For the T-Wing vehicle which is essentially symmetric about both the xz plane (like most aircraft) and the xy plane (unlike most aircraft), $I_{xz} = 0$, and hence many of the inertia terms simplify or disappear.

VERTICAL FLIGHT LINEARIZATION OF EQUATIONS OF MOTION

The nonlinear equations of motion for the T-Wing vehicle, written in body axis form and using vertical Euler angles are given below. The navigation equations have been neglected and the kinematic equations have been converted from the quaternion representation to an Euler angle representation. Also the terms corresponding to zero c_i quantities (because $I_{xz} = 0$) have been dropped.

$$
\dot{u} = rv - qw - g_o \cos(\psi_v)\cos(\theta_v) + \frac{F_x}{m} \tag{7.33}
$$

$$
\dot{v} = -ru + pw + g_o \sin(\psi_v)\cos(\theta_v) + \frac{F_y}{m} \tag{7.34}
$$

$$
\dot{w} = qu - pv - g_o \sin(\theta_v) + \frac{F_z}{m} \tag{7.35}
$$

$$
\dot{p} = c_1\, rq + c_3 L \tag{7.36}
$$

$$
\dot{q} = c_5\, pr + c_7 M \tag{7.37}
$$

$$
\dot{r} = c_8\, pq + c_9 N \tag{7.38}
$$

$$
\dot{\phi}_v = \frac{\cos(\psi_v)}{\cos(\theta_v)}p - \frac{\sin(\psi_v)}{\cos(\theta_v)}q \tag{7.39}
$$

$$
\dot{\theta}_v = \sin(\psi_v)p + \cos(\psi_v)q \tag{7.40}
$$

$$
\dot{\psi}_v = \cos(\psi_v)\tan(\theta_v)p + \sin(\psi_v)\tan(\theta_v)q + r \tag{7.41}
$$

where

$$
c_1 = \frac{(I_{yy} - I_{zz})}{I_{xx}}
$$

$$
c_3 = \frac{1}{I_{xx}}
$$

$$
c_5 = \frac{(I_{zz} - I_{xx})}{I_{yy}}
$$

$$c_7 = \frac{1}{I_{yy}}$$

$$c_8 = \frac{(I_{xx} - I_{yy})}{I_{zz}}$$

$$c_9 = \frac{1}{I_{zz}}$$

To linearize[3] about a given flight state of velocity u_1, the following substitutions are made.

$$
\begin{aligned}
u &\to u_1 + \Delta u \\
v &\to 0 + \Delta v \\
w &\to 0 + \Delta w \\
p &\to 0 + \Delta p \\
q &\to 0 + \Delta q \\
r &\to 0 + \Delta r \\
\phi_v &\to 0 + \Delta \phi \\
\theta_v &\to 0 + \Delta \theta \\
\psi_v &\to 0 + \Delta \psi
\end{aligned}
\tag{7.42}
$$

If equation (7.42) is combined with the equations of motion (7.33)–(7.41), then, using standard small angle assumptions and throwing away all second order terms, a much simpler set of equations is obtained.

$$\dot{u} = -g_o + \frac{F_x}{m} \tag{7.43}$$

$$\dot{v} = -\Delta r \, u_1 + g_o \, \Delta\psi_v + \frac{F_y}{m} \tag{7.44}$$

$$\dot{w} = \Delta q \, u_1 - \Delta\theta_v \, g_o + \frac{F_z}{m} \tag{7.45}$$

$$\dot{p} = \frac{L}{I_{xx}} \tag{7.46}$$

$$\dot{q} = \frac{M}{I_{yy}} \tag{7.47}$$

$$\dot{r} = \frac{N}{I_{zz}} \tag{7.48}$$

$$\dot{\phi}_v = p \tag{7.49}$$

$$\dot{\theta}_v = q \tag{7.50}$$

$$\dot{\psi}_v = r \tag{7.51}$$

The next task is to approximate the force and moment terms using a first order Taylor series expansion. For instance for the F_x forces the expansion is

---

[3] We used the *small disturbances theory*.

$$F_x = F_{x_1} + F_{x_u}\Delta u + F_{x_v}\Delta v + F_{x_w}\Delta w + F_{x_p}\Delta p + F_{x_q}\Delta q + F_{x_r}\Delta r$$
$$+ F_{x_{\delta_e}}\Delta\delta_e + F_{x_{\delta_a}}\Delta\delta_a + F_{x_{\delta_c}}\Delta\delta_c + F_{x_{\delta_r}}\Delta\delta_r + F_{x_{\delta_{th}}}\Delta\delta_{th}$$

In this equation the second subscript denotes the partial derivative with respect to that variable. For instance $F_{x_u} = \frac{\partial F_x}{\partial u}$. The variables δ_e, δ_a, δ_c and δ_r are respectively the elevator, aileron, canard and rudder deflections and δ_{th} is the throttle position. In the case of F_x the zero-order term, F_{x_1}, is non-zero and equal to $-mg_0$. For the force and moment expansions of F_y, F_z, M, L, and N, the zero-order terms are zero.[4]

By neglecting a lot of terms that can be reasonably expected to be close to zero for low-speed vertical flight, it is possible to separate the equations into three sets of *linear differential equations* that are broadly uncoupled from each other. This is in contrast to the normal horizontal flight case, which is typically partitioned into two sets of equations: the longitudinal and lateral cases. The navigation variables have been added back into the sets of variables assuming that the trim state has the vehicle belly (z_b) facing North. In other words the trim values of the vertical Euler angles are all assumed to be zero.

Axial partition

$\mathbf{x}_a = [u, p, \phi_v, h]^T$; $\bar{\mathbf{u}}_a = [\delta_a, \delta_{th}]^T$:

$$\begin{bmatrix} \dot{u} \\ \dot{p} \\ \dot{\phi}_v \\ \dot{h} \end{bmatrix} = \begin{bmatrix} F_{x_u}/m & 0 & 0 & 0 \\ 0 & L_p/I_{xx} & 0 & 0 \\ 0 & 1 & 0 & 0 \\ 1 & 0 & 0 & 0 \end{bmatrix} \begin{bmatrix} u \\ p \\ \phi_v \\ h \end{bmatrix} + \begin{bmatrix} 0 & F_{x_{\delta_{th}}}/m \\ L_{\delta_a}/I_{xx} & 0 \\ 0 & 0 \\ 0 & 0 \end{bmatrix} \begin{bmatrix} \delta_a \\ \delta_{th} \end{bmatrix} \quad (7.52)$$

Longitudinal partition

$\mathbf{x}_{long} = [w, q, \theta_v, p_N]^T$; $\bar{\mathbf{u}}_{long} = [\delta_e]$:

$$\begin{bmatrix} \dot{w} \\ \dot{q} \\ \dot{\theta}_v \\ \dot{p_N} \end{bmatrix} = \begin{bmatrix} F_{z_w}/m & -u_1 + F_{z_q}/m & -g_0 & 0 \\ M_w/I_{yy} & M_q/I_{yy} & 0 & 0 \\ 0 & 1 & 0 & 0 \\ 1 & 0 & -u_1 & 0 \end{bmatrix} \begin{bmatrix} w \\ q \\ \theta_v \\ p_N \end{bmatrix} + \begin{bmatrix} F_{z_{\delta_e}}/m \\ M_{\delta_e}/I_{yy} \\ 0 \\ 0 \end{bmatrix} \delta_e \quad (7.53)$$

---

[4] These zero-order terms are only applicable for purely vertical non-accelerating flight.

Lateral partition

$\mathbf{x}_{lat} = [v, r, \psi_v, p_E]; \; \bar{\mathbf{u}}_{lat} = [\delta_r]^T:$

$$
\begin{bmatrix} \dot{v} \\ \dot{r} \\ \dot{\psi}_v \\ \dot{p}_E \end{bmatrix} = \begin{bmatrix} F_{y_v}/m & -u_1 + F_{y_r}/m & g_0 & 0 \\ N_v/I_{zz} & N_r/I_{zz} & 0 & 0 \\ 0 & 1 & 0 & 0 \\ 1 & 0 & u_1 & 0 \end{bmatrix} \begin{bmatrix} v \\ r \\ \psi_v \\ p_E \end{bmatrix} + \begin{bmatrix} F_{y_{\delta_r}}/m \\ N_{\delta_r}/I_{zz} \\ 0 \\ 0 \end{bmatrix} \delta_r \qquad (7.54)
$$

7.7 Real-time Flight Simulation

Using the aerodynamic model previously described a full nonlinear 6-degree of freedom (DOF) flight simulation of the actual technology demonstrator vehicle has been constructed. This enables the prediction of the dynamics of the vehicle to be calculated in all phases of flight from hover, through the transition manoeuvre to full-speed horizontal flight and includes real-world effects such as wind gusts and sensor errors. This model is currently implemented in the SIMULINK [149] environment and has now been refined to run in real time via use of an add-on product, the Real Time Workshop [143]. This is a significant advance on the previous simulation work, which ran at approximately 1/30 th of actual speed.

One of the initial uses of this was a visual pilot training simulator, which allowed a ground-based remote pilot to practice manual flight of the vehicle without risking an airframe. This was required because the technology demonstrator vehicle was initially test flown without any automatic controls in place. More recently it has also been the basis for the development of stability augmentation systems (SAS) and then full automatic controls. By coupling models of these control systems with the basic flight dynamic SIMULINK model and then re-compiling the real-time executive it has also been possible to test new controller designs quickly and easily. This is particularly true for SAS controllers: new designs can be incorporated and flown in real-time on the simulation by a pilot to gauge how well they perform. The use of control-system rapid prototyping tools such as the Real Time Workshop and xPCTarget has substantially compressed the control system design cycle.

The improved vehicle simulation has also been important in optimizing the transition manoeuvres for the vehicle. One of the goals of the current programme is to make these manoeuvres as efficient as possible. In practical terms this means minimizing the altitude loss in the vertical to horizontal transition, while minimizing the altitude gain for the reverse manoeuvre. In performing these optimizations a fast simulation is of great benefit.

A figure showing snapshots of a typical vehicle undergoing a vertical to horizontal transition manoeuvre under rudimentary automatic control is given in Figure 7.11. Figure 7.12 shows the flight simulation.

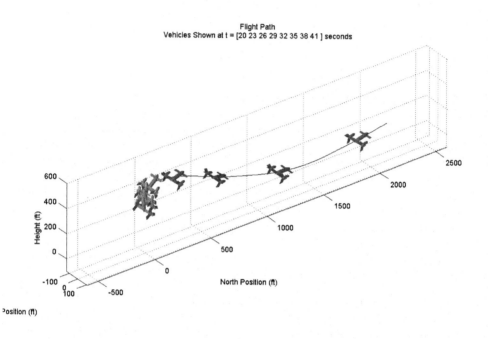

Fig. 7.11. Typical simulation of vertical to horizontal transition under automatic control.

7.8 Hover Control Model

The hover control of the vehicle in the presence of wind gusts and the maximum control authority in the presence of steady winds are critical features of the tail-sitter design. The vertical flight controllability and control authority analysis is based on the aerodynamic coefficients (control and damping derivatives) obtained from the panel method analysis coupled with a mass model of the vehicle. By linearizing and partitioning the vehicle hover dynamics it is possible to automate the design of simple LQR (Linear Quadratic Regulator) controllers to stabilize the vehicle in the presence of prescribed wind gusts and incorporate this directly into the optimization problem.

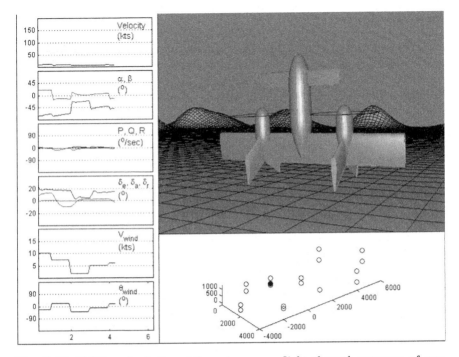

Fig. 7.12. T-Wing simulation: fully autonomous flight through sequence of way-points.

7.8.1 Vertical Flight Controllers

For vertical flight of the T-Wing vehicle, the low-level controllers used are effectively velocity controllers. Specifically they are:

- W – velocity elevator controller. This controls the velocity component normal to the belly of the vehicle, using the elevons as elevators.

- V – velocity rudder controller. This controls the velocity component in the body-axis y-direction, via the rudders.

- P – vertical roll rate controller. Control in this case is effected via the elevons acting as ailerons. This controls the roll-rate of the vehicle about its longitudinal x-axis, which will be predominantly vertical for typical vertical-mode flight.

- $\dot{h}$ – vertical-velocity throttle controller. This controls vertical velocity via changes in engine throttle setting.

Controller design

In the context of dealing with horizontal wind gusts, only the longitudinal and lateral states are significant; the axial states are less subject to disturbance and relatively easy to control. In the following, for reasons outlined above, only the longitudinal dynamics and the design of a W-controller will be considered. Taking the equations for the longitudinal case (7.53), a further simplification can be made if it is assumed that the vehicle is in a pure hover, with zero upward speed. This is shown in the following equation, which forms the basis for the control design process.

$$
\begin{bmatrix} \dot{w} \\ \dot{q} \\ \dot{\theta}_v \\ \dot{p}_N \end{bmatrix} = \begin{bmatrix} F_{z_w}/m & F_{z_q}/m & -g_0 & 0 \\ M_w/I_{yy} & M_q/I_{yy} & 0 & 0 \\ 0 & 1 & 0 & 0 \\ 1 & 0 & 0 & 0 \end{bmatrix} \begin{bmatrix} w \\ q \\ \theta_v \\ p_N \end{bmatrix} + \begin{bmatrix} F_{z_{\delta_e}}/m \\ M_{\delta_e}/I_{yy} \\ 0 \\ 0 \end{bmatrix} \delta_e \qquad (7.55)
$$

This equation can be used by itself to develop full LQR position controllers for the vehicle. An alternative is to use it to develop velocity controllers in which case the p_N variable is re-interpreted as simply the integral of the W-velocity component, I_W (irrespective of vehicle orientation) thus allowing for integral action within the controller. Doing this ensures zero-steady-state error properties for the controllers so designed.

Using the system given in equation (7.55), the controller design can now be performed as indicated below. The system has the standard state space form:

$$
\dot{\mathbf{x}} = \mathbf{A}\mathbf{x} + \mathbf{B}\mathbf{u} \qquad (7.56)
$$

where

$$
\mathbf{x} = [w \ q \ \theta_v \ I_W]^T; \qquad \mathbf{u} = \delta_e
$$

$$
\mathbf{A} = \begin{bmatrix} -1.309 & 0.039 & -32.2 & 0 \\ -0.353 & -0.626 & 0 & 0 \\ 0 & 1 & 0 & 0 \\ 1 & 0 & 0 & 0 \end{bmatrix}
$$

$$
\mathbf{B} = \begin{bmatrix} -0.278 \\ -0.277 \\ 0 \\ 0 \end{bmatrix}
$$

$$(7.57)$$

To perform an LQR design, all that is required is to select the $\mathbf{Q}$-matrix that penalizes state errors and the $\mathbf{R}$-matrix that penalizes control displacements.

Using these design matrices the MATLAB $care()$ function can be invoked to perform the solution of the algebraic Ricatti equation, to find optimal gains to minimize the combined state-deviation and control action errors. The values for the gain matrix are

$$\mathbf{Q} = diag(1, 0, 10000, 5); \qquad \mathbf{R} = 10$$

$$\mathbf{K} = 2.13 \quad -18.3 \quad -52.3 \quad 0.707$$

(7.58)

The control input can be written as

$$\delta_e = 2.13(w_{command} - w) + 18.3\, q + 52.3\, \theta_v$$
$$+ 0.707 \int (w_{command} - w)dt \ (°)$$

(7.59)

The output for this control law is shown in Figure 7.13. This controller has a fast response but a small overshoot ($\sim$ 18%). One advantage of this controller is that its steady-state error properties are guaranteed by the structure of the controller due to the integral term.

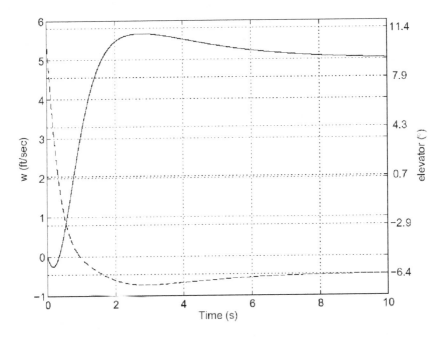

Fig. 7.13. w-Velocity controller with integral action: w (solid) and elevator (dashed). Response to 5 ft/sec step input.

The gain-scheduling of the v-controllers for vertical translational flight was done in an exactly similar way to that of the w-controllers.

Roll-rate or "p" controller design

The control of vertical roll-rate, p, is accomplished via the use of a classical compensator to control ailerons based on the error between the actual and commanded roll-rate of the vehicle. The block diagram implementing the simple proportional controller is shown in Figure 7.14. The step-response of this compensator is shown in Figure 7.15.

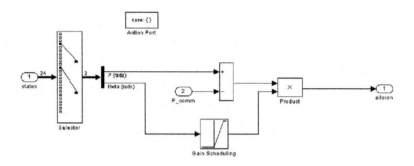

Fig. 7.14. Block diagram of p-controller. Note gain scheduling based on pitch-angle (gain scheduling based on theta because we cannot assume we always know velocity).

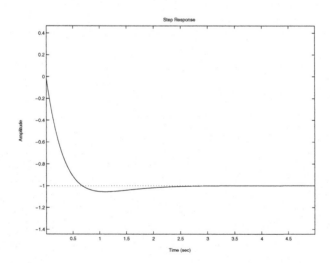

Fig. 7.15. Roll-rate controller for vertical flight: step response (not currently implemented).

Throttle vertical flight $\dot{h}$ controller

The throttle controller for vertical flight is a simple direct gain controller, with a simple trim throttle offset. The trim offset is set at 0.56 (56%), the value required for ascending vertical flight at 1 ft/sec. A low gain of 5% throttle/(ft/sec) on the velocity error has been selected to prevent continual throttle over-reaction to sensed changes in velocity.

A saturation block with a lower limit of 48% has been used on the throttle output in vertical flight to prevent the throttle ever going too low. This is of vital importance for overall control of the vehicle as all the other controllers depend on there being sufficient dynamic pressure provided by the propeller slipstream to generate control forces. It should also be noted that if the vehicle empty weight is changed significantly then this lower limit may have to be re-evaluated to enable the vehicle to initiate and maintain descending vertical flight. However, for the current vehicle the lower throttle limit of 48% seems a reasonable one. In fact better vertical flight control may be achieved if this value is increased slightly (to say 50%). Poorer performance will be observed if the value is reduced to 40%. A picture of the throttle-controller block diagram is given in Figure 7.16.

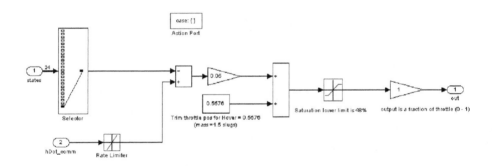

Fig. 7.16. Throttle vertical ($\dot{h}$) controller.

7.9 Flight of T-Wing Vehicle

Fig. 7.17. First manual hover flight of T-Wing (11 December 2000).

Fig. 7.18. Top view of the T-Wing aircraft (August 2002).

Fig. 7.19. First autonomous flight of T-Wing aircraft in September 2002.

7.10 Conclusion

In conclusion it can be seen that the tail-sitter UAV concept being developed by Sydney University and Sonacom exhibits great promise both for defence and civilian applications in the future. Progress from purely theoretical analyses to physical hardware has now occurred. The programme is currently entering a significant period of flight-testing during which the level of vehicle autonomy will be steadily increased, culminating in a full autonomous flight from take-off to landing. It is the author's belief that this UAV implementation will finally allow the tail-sitter concept to realize its full potential: something that was never possible for manned tail-sitter vehicles.

8

Modelling and Control of Small Autonomous Airships

Dr. Y. Bestaoui[1]
University of Evry, France

8.1 Introduction

Lighter than air vehicles suit a wide range of applications, ranging from advertising, aerial photography and aerial inspection platforms, with a very important application area in environmental, biodiversity, and climatological research and monitoring. Airships[2] offer the advantage of quiet hover with noise levels much lower than helicopters. Unmanned remotely-operated airships have already proved themselves as camera and TV platforms and for specialized scientific tasks. An actual trend is toward autonomous airships. Such vehicles must be able to hover above an area and must have extended airborne capabilities for long-duration flights. They must generate very low noise and turbulence as well as very low vibrations. Significant advances in motion planning and control must be made in order for an airship to be autonomous. The current flight operation can be: take-off, cruise, turn, hover, land, etc.

Figure 8.1 shows the Laboratoire des Systemes Complexes (LSC) airship AS200. It is basically a large gas balloon. Its shape is maintained by its internal overpressure. The only solid parts are the gondola, a tail rotor with horizontal axis of rotation, the main rotor with a varying tilt angle and the tail fins. It is actually a remotely piloted airship designed for remote sensing, being transformed to be fully autonomous. In the literature, mostly classical linear

---

[1] This chapter was written in collaboration with Dr. L. Beji and Dr. A. Abichou.
[2] An airship is a buoyant ("lighter-than-air") aircraft that can be steered and propelled through the air. Airships are also known as dirigibles. Blimp is an informal term typically applied to non-rigid, helium-filled, airships.

control, decoupling longitudinal and lateral modes, is investigated [27, 38, 83]. For the purpose of studying the stability and control of the airship, the motion of the axes is considered with respect to an initial condition. When it is assumed that motion of the airship is constrained to small perturbations about the trimmed equilibrium flight condition then the model may be considerably simplified. In particular, the products and squares of small perturbation variables become negligibly small and since the attitude perturbations are also small, their sines and cosines may take small angle approximations. In small perturbations, it is also reasonable to assume decoupled longitudinal and lateral motion and consequently all coupling derivatives may be omitted from the equations. The airship dynamics are divided into fast (short period) and slow (phugoid) motions.

However, some attempts at nonlinear control have also been made in [13, 14, 18, 132], using the full dynamics model. Airships offer a control challenge as they have non-zero drift and their linearization at zero velocity is not controllable.

Fig. 8.1. LSC airship AS200.

In this chapter, we consider the stabilizing control problem using only the three available inputs: the main and tail thrusters and the tilt angle of the main propeller. The roll is totally unactuated. The same input controls both pitch and surge, while yaw and sway are related. The unactuated dynamic implies constraints on the accelerations. The dynamic positioning control problem consists of finding a feedback control law that asymptotically stabilizes both position and orientation to fixed constant values. Most small time locally controllable systems can be stabilized by continuous time-varying feedback. Homogeneous approximations and feedback have been applied successfully to systems without drift [109] and to some systems with drift vector [134]. In fact, no necessary and sufficient conditions have been found for small time controllable systems. Reference [134] has recently proposed a model for an underwater vehicle which has a homogeneous structure. Their model also fails Brockett's condition for continuous time invariant stabilization. The method used is basically a case by case study.

In this work, a periodic time-varying feedback law is developed. The feedback control law is derived using averaging theory and homogeneity properties. It is based on a quaternion representation of the orientation. We prove that it stabilizes asymptotically both the position and orientation of the airship using only the three available control inputs. Averaging is an important tool used in the analysis of time-varying systems [14, 13]. An auxiliary time-invariant dynamical system

$$\dot{x} = f_{av}(x) \tag{8.1}$$

called the average, is used to investigate properties of a time-varying dynamical system

$$\dot{x} = f\left(\frac{t}{\varepsilon}, x\right) \tag{8.2}$$

that depends on a small parameter ε. The search for locally exponentially stabilizing feedback has been reduced to looking for time-periodic asymptotically stabilizing homogeneous degree-one functions for the approximate system.

8.2 Euler–Lagrange Modelling

8.2.1 Kinematics

Consider a rigid body moving in free space (Figure 8.2). Two reference frames are considered in the derivation of the kinematics and dynamic equations of motion. These are the earth fixed frame R_f and the body fixed frame R_m.

The position and orientation of the vehicle should be described relative to the inertial reference frame while the linear and angular velocities of the vehicle should be expressed in the body-fixed coordinate system.

The origin C of R_m coincides with the centre of gravity (cg) of the vehicle. Its axes (X_V, Y_V, Z_V) are the principal axes of symmetry when available. They must form a right-handed orthogonal normed frame. The position of the vehicle C in R_f can be described by

$$\eta_1 = \begin{pmatrix} x \\ y \\ z \end{pmatrix} \tag{8.3}$$

At each instant, the configuration (position and orientation) of the rigid body can be described by a homogeneous transformation matrix corresponding to the displacement from frame R_f to frame R_m.

The set of all such matrices is called $SE(3)$, the special Euclidean group of rigid body transformations in 3D.

$$SE(3) = \left\{ A \middle| A = \begin{pmatrix} R & d \\ 0 & 1 \end{pmatrix}, R \in \mathbb{R}^{3\times3}, d \in \mathbb{R}^3, R^T R = I, \det(R) = 1 \right\} \tag{8.4}$$

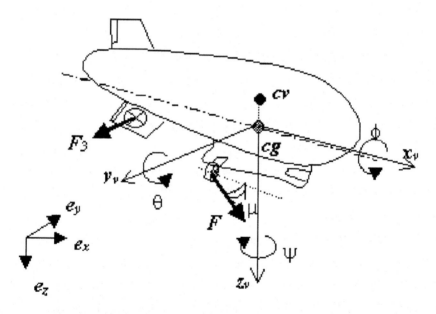

Fig. 8.2. Geometric representation of an airship.

$SE(3)$ is a Lie group for the standard matrix multiplication and it is a manifold. On any Lie group, the tangent space at the group identity has the structure of a Lie algebra. The Lie algebra of $SE(3)$ denoted by $se(3)$, is given by

$$se(3) = \left\{ \begin{pmatrix} \hat{\omega} & V \\ 0 & 0 \end{pmatrix}, \hat{\omega} \in \mathbb{R}^{3\times3}, V \in \mathbb{R}^3, \hat{\omega}^T = -\hat{\omega} \right\} \tag{8.5}$$

A 3×3 skew symmetric matrix $\hat{\omega}$ can be identified with a vector $\omega \in \mathbb{R}^3$ so that for any arbitrary vector $x \in \mathbb{R}^3$, $\hat{\omega}x = \omega * x$ where $\omega * x$ is the vector cross-product operation in $\mathbb{R}^3$ and

$$\hat{\omega} = \begin{pmatrix} 0 & -\omega^3 & \omega^2 \\ \omega^3 & 0 & -\omega^1 \\ -\omega^2 & \omega^1 & 0 \end{pmatrix} \tag{8.6}$$

Each element $S \in se(3)$ can thus be identified with a vector pair $\{\omega\ V\}$. Given a curve $C(t) : [-a\ a] \to SE(3)$, an element $S(t)$ of the Lie algebra $se(3)$ can be associated to the tangent vector $\dot{C}(t)$ at an arbitrary point t by

$$SE(3) = A^{-1}(t)\dot{A}(t) = \begin{pmatrix} \hat{\omega} & R^T \dot{d} \\ 0 & 0 \end{pmatrix} \tag{8.7}$$

where $\hat{\omega}(t) = R^T(t)\dot{R}(t)$ is the corresponding element of $SO(3)$.

A curve on $SE(3)$ physically represents a motion of the rigid body. If $\{\omega(t)\ V(t)\}$ is the pair corresponding to $S(t)$, then ω physically corresponds to the angular velocity of the rigid body, while V is the linear velocity of the origin O' of the frame $\{M\}$ with

$$V = \begin{pmatrix} u \\ v \\ w \end{pmatrix} \tag{8.8}$$

$$\omega = \begin{pmatrix} p \\ q \\ r \end{pmatrix} \tag{8.9}$$

To avoid the singularity inherent to the representation of Euler angles, we have chosen Euler parameters as the orientation representation. The Euler parameters are defined as a unit quaternion and are represented by a point on the surface of the 4-dimensional unit hypersphere S^3. Some authors call them Cayley–Klein parameters [52]. The Euler parameters are expressed by the rotation axis n and the rotation angle β about the axis as follows:

$$\eta_2 = \begin{pmatrix} \cos\left(\frac{\beta}{2}\right) \\ \sin\left(\frac{\beta}{2}\right) n \end{pmatrix} = \begin{pmatrix} e_0 \\ e_1 \\ e_2 \\ e_3 \end{pmatrix} \tag{8.10}$$

$$R(\eta_2) = \begin{pmatrix} 1 - 2(e_2^2 + e_3^2) & 2(e_1e_2 - e_3e_0) & 2(e_1e_3 + e_2e_0) \\ 2(e_1e_2 + e_3e_0) & 1 - 2(e_1^2 + e_3^2) & 2(e_2e_3 - e_1e_0) \\ 2(e_1e_3 - e_2e_0) & 2(e_2e_3 + e_1e_0) & 1 - 2(e_1^2 + e_2^2) \end{pmatrix} \tag{8.11}$$

$$0 \le \beta \le 2\pi$$

where

$$e = \sin\left(\frac{\beta}{2}\right) n \qquad (8.12)$$

The associated kinematical differential equation of Euler parameters is given by

$$\dot{e} = \frac{1}{2}\left(e_0 I_{3\cdot3} + \hat{e}\right) \omega \qquad (8.13)$$

$$\dot{e}_0 = -\frac{1}{2} e^T \omega \qquad (8.14)$$

and define a vector field on S^3, the 3-sphere on $\mathbb{R}^4$. $\mathbb{R}^4$ represents the set of 4×1 real vectors. The kinematics of the airship can be expressed in the following way:

$$\begin{pmatrix} \dot{\eta}_1 \\ \dot{\eta}_2 \end{pmatrix} = \begin{pmatrix} R & 0_{3\cdot3} \\ 0_{3\cdot3} & J(\eta_2) \end{pmatrix} \begin{pmatrix} V \\ \omega \end{pmatrix} \qquad (8.15)$$

where

$$J(\eta_2) = \frac{1}{2} \begin{pmatrix} -e_1 & -e_2 & -e_3 \\ e_0 & -e_3 & e_2 \\ e_3 & e_0 & -e_1 \\ -e_2 & e_1 & e_0 \end{pmatrix} \qquad (8.16)$$

and

$$J^T(\eta_2)J(\eta_2) = \frac{1}{4} I_{3\cdot3} \qquad (8.17)$$

We choose a parametrization of $SE(3)$ induced by the product structure $SO(3) \times \mathcal{R}^3$. The set of all positions and orientations being not Euclidean, there is no obvious choice of a metric on this set. The norm of the velocity and the distance between two positions and orientations is not defined. Through the transformation between the rotation matrix and the quaternion, the modelling and the analysis can be performed in terms of the rotation matrix while the quaternion representation is used in the control design procedure.

8.2.2 Dynamics

In this section, analytic expressions for the forces and moments of a system with added mass and inertia such as an airship are introduced [16, 177, 52]. An airship is a lighter-than-air vehicle using a lifting gas (helium in this particular case). Since the volume of the gondola is negligible compared with that of the envelope, it is reasonable to assume that the centre of volume (cv) lies on the axis of symmetry of the envelope. The buoyancy force B acts at the centre of buoyancy (cb) and the weight acts at the centre of gravity. The total engine thrust acts at a point below the Ox axis but its precise location depends on the geometry of the installed propulsion system. We will make in the sequel some simplifying assumptions: the earth fixed frame is inertial, the gravitational field is constant, the airship is well inflated, the density of air is uniform. [11] consider the case of an airship with small deformations analysed via the updated Lagrangian method.

The translational part being separated from the rotational part [16], the dynamic equations (Euler–Poincaré) are given by

$$MV̇ = -\omega * MV - b(.) + f(u) \tag{8.18}$$
$$J\dot{\omega} = -\omega * J\omega - V * MV - \gamma(.) + \tau(u) \tag{8.19}$$

where M and J are respectively the vehicle's mass and rotational tensors and τ, γ and f, b represent respectively the control and non-conservative torques and forces in body axes. For a system with added masses, the term $V * MV$ is non-zero. We can propose

$$b(.) = R^T e_z(mg - B) + (D_V)V \tag{8.20}$$
$$\gamma(.) = (R^T e_z * \overline{BG})B + (D_\omega)\omega \tag{8.21}$$

where m is the mass of the airship, the propellers and actuators. M includes both the airship's actual mass as well as the virtual mass elements associated with the dynamics of buoyant vehicles. J includes both the airship's actual inertias as well as the virtual inertia elements associated with the dynamics of buoyant vehicles. As the airship displays a very large volume, its added masses and inertias become very significant [52]. We will assume that the added mass coefficients are constant. They can be estimated from the inertia ratios and the airship weight and dimension parameters. Added mass should be understood as pressure-induced forces and moments due to harmonic motion of the body which are proportional to the acceleration of the body. In order to allow the vehicle to pass through the air, the fluid must move aside and then close behind the vehicle. As a consequence, the fluid passage possesses kinetic energy due to the motion of the vehicle [52].

We also have that Be_z is a 3×1 buoyancy force vector, and $B = \rho \Delta g$ where Δ is the volume of the envelope, ρ is the difference between the density of the ambient atmosphere ρ_{air} and the density of the helium ρ_{helium} in the envelope, g is the constant gravity acceleration and $e_z = (0, 0, 1)^T$ a unit vector.

The aerodynamic force can be divided into two component forces, one parallel and the other perpendicular to the direction of motion. Lift is the component of the aerodynamic force perpendicular to the direction of motion and drag is the component opposite to the direction of motion. As the airship is a slow-moving object in the air, we can assume a linear relationship between the speed and the drag. D_V is the 3×3 aerodynamic forces diagonal matrix and D_ω is the 3×3 aerodynamic moments diagonal matrix:

$$D_V = diag(-X_u \ -Y_v \ -Z_w) \tag{8.22}$$
$$D_\omega = diag(-L_p \ -M_q \ -N_r) \tag{8.23}$$

The gravitational force vector is given by the difference between the airship weight and the buoyancy acting upwards on it:

$$R^T e_z(mg - B) = (mg - B) \begin{pmatrix} 2(e_1e_3 - e_2e_0) \\ 2(e_2e_3 + e_1e_0) \\ 1 - 2(e_1^2 + e_2^2) \end{pmatrix} \tag{8.24}$$

The gravitational moments are given by

$$(R^T e_z * \overline{BG})B = B \begin{pmatrix} 2z_b(e_2e_3 + e_1e_0) - y_b(1 - 2(e_1^2 + e_2^2)) \\ x_b(1 - 2(e_1^2 + e_2^2)) + z_b(e_1e_3 - e_2e_0) \\ 2y_b(e_1e_3 - e_2e_0) - 2x_b(e_2e_3 + e_1e_0) \end{pmatrix} \tag{8.25}$$

where $\overline{BG} = (x_b \ y_b \ z_b)$ represents the position of the centre of buoyancy with respect to the body fixed frame.

If a system is fully actuated, it can be steered along any given curve on the configuration manifold, i.e. it is controllable. This is not true in general for an underactuated system. However, an underactuated system can be locally controllable if it enjoys the property of nonholonomy. The existence of non-holonomic constraints translates into the fact that the system can be locally steered along a manifold of dimension larger than the number of independent control inputs [126, 17, 50, 67, 68]. However, practical methods have only been found for simple underactuated systems.

8.2.3 Propulsion

Actuators provide the means for manoeuvring the airship along its course. An airship is propelled by thrust. Propellers are designed to exert thrust to drive the airship forward. The most popular propulsion system layout for pressurized non-rigid airships is twin ducted propellers mounted either side of the envelope bottom. Another one exists in the tail for torque correction and attitude control. In aerostatics hovering (floating), its stability is mainly affected by its centre of buoyancy in relation to the centre of gravity. The airship's centre of gravity can be adjusted to obtain either stable, neutral or unstable conditions. Putting all the weight on the top would create a highly unstable airship with a tendency to roll over in a stable position.

In aerodynamical flight, stability can be affected by fins and the general layout of the envelope. Control inertia can be affected by weight distribution, dynamic (static) stability and control power (leverage) available.

The airship AS200 is an underactuated system with two types of control in a low-velocity flight: forces generated by thrusters and angular inputs controlling the direction of the thrusters (γ is the tilt angle of the propellers):

$$f(u) = F_1 + F_2 \tag{8.26}$$

$$\tau(u) = -F_1 * \overline{P_1G} - F_2 * \overline{P_2G} \tag{8.27}$$

where

$$F_1 = \begin{pmatrix} T_M \sin(\gamma) \\ 0 \\ T_M \cos(\gamma) \end{pmatrix} \tag{8.28}$$

$$F_2 = \begin{pmatrix} 0 \\ T_T \\ 0 \end{pmatrix} \tag{8.29}$$

and T_M, T_T represent respectively the main and tail thrusters. If we consider the plane $X - Z$ as a plane of symmetry, the mass and inertia matrices can be written as

$$M = \begin{pmatrix} m + X_u & 0 & X_z \\ 0 & m + Y_v & 0 \\ Z_x & 0 & m + Z_w \end{pmatrix} \tag{8.30}$$

$$J = \begin{pmatrix} I_x + L_p & 0 & -I_{xz} \\ 0 & I_y + M_q & 0 \\ -I_{xz} & 0 & I_z + N_r \end{pmatrix} \tag{8.31}$$

If the centre of gravity sits below the centre of buoyancy, then $\overline{BG} = (0\ 0\ Z_b)^T$.

Thus in building the nonlinear six degrees of freedom mathematical model, the additional following assumptions are made:

$$\overline{P_1G} = \begin{pmatrix} 0 \\ 0 \\ P_1^3 \end{pmatrix} \tag{8.32}$$

$$\overline{P_2G} = \begin{pmatrix} P_2^1 \\ 0 \\ 0 \end{pmatrix} \tag{8.33}$$

It is important to gain insight into the geometric structure of the equations since this knowledge can be useful in areas such as motion planning and control.

8.3 Stabilization Problem

With the assumptions stated above, the dynamics and kinematics of a small airship can be written in the following compact form:

$$M_v \dot{v} + C_v(v)v + g_v(\eta_2) = B_\tau \tau \tag{8.34}$$

$$\dot{\eta} = J(\eta_2)v \tag{8.35}$$

$$v = (V \ w)^T \tag{8.36}$$

where $M_v \in \mathbb{R}^{6\times6}$ is the inertia matrix which is block diagonal and a constant matrix (symmetric and positive definite):

$$M_v = \begin{pmatrix} m_{11} & 0 & 0 & 0 & 0 & 0 \\ 0 & m_{22} & 0 & 0 & 0 & 0 \\ 0 & 0 & m_{33} & 0 & 0 & 0 \\ 0 & 0 & 0 & I_{11} & 0 & I_{13} \\ 0 & 0 & 0 & 0 & I_{22} & 0 \\ 0 & 0 & 0 & I_{13} & 0 & I_{33} \end{pmatrix} \tag{8.37}$$

The centrifugal and Coriolis terms $(C_v(v) \in R^{6\times6})$ can be rewritten in the form

$$C_v(v) = \begin{pmatrix} C_v^1 & 0_{3\times3} \\ C_v^2 & C_v^3 \end{pmatrix} \tag{8.38}$$

where

$$C_v^1 = \begin{pmatrix} 0 & -m_{22}r & m_{33}q \\ m_{11}r & 0 & -m_{33}p \\ -m_{11}q & m_{22}p & 0 \end{pmatrix} \tag{8.39}$$

$$C_v^2 = \begin{pmatrix} 0 & (M_{22} - M_{33})w & 0 \\ 0 & 0 & (M_{33} - M_{11})u \\ (M_{11} - M_{22})v & 0 & 0 \end{pmatrix} \tag{8.40}$$

$$C_v^3 = \begin{pmatrix} I_{13}q & -I_{22}r & I_{33}q \\ -I_{13}p - I_{11}r & 0 & I_{33}p + I_{13}r \\ -I_{11}q & I_{22}p & -I_{13}q \end{pmatrix} \tag{8.41}$$

The constant positive definite damping matrix $(D_v(v) \in R^{6\times6})$ takes the form

$$D_v(v) = \text{diag}(D_v, D_\Omega) \tag{8.42}$$

The gravitational vector $(g_v(\eta_2) \in R^6)$ is defined by

$$g_v(\eta_2) = \begin{pmatrix} 2(B - mg)(e_1e_3 - e_0e_2) \\ 2(B - mg)(e_2e_3 + e_0e_1) \\ (B - mg)(1 - 2(e_1^2 + e_2^2)) \\ -2Bz_b(e_0e_1 + e_2e_3) \\ 2Bz_b(e_0e_2 - e_1e_3) \\ 0 \end{pmatrix} \tag{8.43}$$

and the constant matrix

$$B_\tau = \begin{pmatrix} 1 & 0 & 0 \\ 0 & 1 & 0 \\ 0 & 0 & 1 \\ 0 & 0 & 0 \\ P_1^3 & 0 & 0 \\ 0 & -P_2^1 & 0 \end{pmatrix} \tag{8.44}$$

In the sequel, the control input $(\tau \in R^3)$ will be taken as

$$\tau = \begin{pmatrix} \tau_1 \\ \tau_2 \\ \tau_3 \end{pmatrix} = \begin{pmatrix} T_M \sin(\mu) \\ T_T \\ T_M \cos(\mu) \end{pmatrix} \tag{8.45}$$

Notice that the transformation $(\tau_1\ \tau_2\ \tau_3) \to (T_M\ \mu\ T_T)$ is a diffeomorphism. Then we can select $(\tau_1\ \tau_2\ \tau_3)$ as a control vector for the airship.

We will show first that it is not possible to stabilize the airship using a feedback law that is a continuous function of the state only. This follows from results by [23, 28, 29, 109]. The problem is thus not solvable using linearization and linear control theory or classical nonlinear control theory like feedback linearization. Thus, we propose a continuous periodic time-varying feedback law that stabilizes the airship using only the three available inputs.

Proposition 8.1. *The system cannot be stabilized by a continuous pure state feedback law.*

Proof. Let us consider $\varepsilon = (\varepsilon_1\ 0)^T$; we will have $v = 0$ since $JJ^T = \frac{1}{4}I_{4\times4}$. Therefore equation (8.37) leads to

$$B_\tau \tau - g(\eta_2) = M_v \varepsilon_1 \tag{8.46}$$

Then if we take $\varepsilon_1 = (0\ \varepsilon_0\ 0\ 0\ 0)^T$ with $\varepsilon_0 \neq 0$, we will obtain the following system

$$\tau_1 - 2(B - mg)(e_1 e_3 - e_0 e_2) = 0 \tag{8.47}$$
$$\tau_2 - 2(B - mg)(e_2 e_3 + e_0 e_1) = m_{22}\epsilon_0 \tag{8.48}$$
$$\tau_3 - (B - mg)(1 - 2(e_1^2 + e_2^2)) = 0 \tag{8.49}$$
$$2Bz_b(e_0 e_1 + e_2 e_3) = 0 \tag{8.50}$$
$$P_1^3 \tau_1 + 2Bz_b(e_3 e_1 - e_2 e_0) = 0 \tag{8.51}$$
$$-P_2^1 \tau_2 = 0 \tag{8.52}$$

We can deduce from the last equations that $\tau_2 = 0$. Furthermore, the fourth equation implies that $e_0 e_1 + e_2 e_3 = 0$. As a result, $m_{22}\varepsilon_0 = 0$, which is impossible since $\varepsilon_0 \neq 0$. Therefore, we cannot stabilize the airship using a continuous pure-state feedback (Brockett's necessary condition [23]). However, Coron's theorem [29] proves that time-periodic continuous feedback is sufficient to stabilize the system to a point.   □

Instead of working with the original inputs, an approximation that makes sense in terms of stabilization about a reference point is sought. The Jacobian linearization of this system about any point is not useful in any control-theoretic context since the linearized system is not controllable. However, if the Lie algebra of the set of analytic input vector fields has rank n then there exists a homogeneous approximate system of degree one. The use of homogeneous feedback is strongly motivated by the existence of a controllable homogeneous approximate system. A homogeneous degree one control function should be found such that the origin is uniformly asymptotically stable.

In the sequel, we develop a continuous time-varying feedback law. The main result is given by the following proposition.

Proposition 8.2. *Consider the function*

$$p_d = -k^r r - k^{e_3} e_3 - k^{e_1} e_1 + \frac{k^v v + k^y y}{\sqrt{|v| + |y|}} \sin\left(\frac{t}{\varepsilon}\right) \tag{8.53}$$

$$w_d = -k^z e_z + 2\sqrt{|v| + |y|} \sin\left(\frac{t}{\varepsilon}\right) \tag{8.54}$$

$$q_d = -k^{e_2} e_2 - k^x x - k^u u \tag{8.55}$$

Furthermore, consider the following time-varying continuous feedback

$$\tau_1(v, \eta, t) = \frac{1}{P_1^3}\left((I_{22}k^q - M_q)q - I_{22}k^q q_d + 2B z_b e_2\right) \tag{8.56}$$

$$\tau_2(v, \eta, t) = \frac{1}{P_{12}^1 I_{13}}\left(-(\Delta k^p - L_p I_{33})p + \Delta k^p p_d - 2B z_b e_1 I_{33}\right) + \frac{N_r}{P_2^1}r \tag{8.57}$$

$$\tau_3(v, \eta, t) = (m_{33}k^w - Z_w)w - m_{33}k^w w_d + B - mg \tag{8.58}$$

with $\Delta = I_{13}^2 - I_{11}I_{33}$. Then with a suitable choice of the positive parameters $k^r, k^{e_3}, k^{e_2}, k^{e_1}, k^z, k^x, k^u$ there exists ε_0 such that for any $\varepsilon \in (0, \varepsilon_0]$ and large enough k^q, k^p, k^w, the feedback stabilizes locally exponentially the system. ε is a parameter we need to adjust.

Proof. Let consider the following dilation:

$$x_\lambda^\alpha(v, \eta, t) = \left(\lambda u, \lambda^2 v, \lambda w, \lambda p, \lambda q, \lambda r, \lambda x, \lambda^2 y, \lambda z, \lambda^2 e_0, \lambda e_1, \lambda e_2, \lambda e_3, t\right) \tag{8.59}$$

The initial system of equation (8.37) can be rewritten as

$$\begin{pmatrix} \dot{v} \\ \dot{\eta} \end{pmatrix} = f(v, \eta, t) + g(v, \eta, t) \tag{8.60}$$

with

$$f(v, \eta, t) = \begin{pmatrix} (X_u u + 2(B - mg)e_2 + \tau_1)/m_{11} \\ (Y_v v - 2(B - mg)e_1 + m_{33}pw + \tau_2)/m_{22} \\ (Z_w w - (B - mg) + \tau_3)/m_{33} \\ (-L_p I_{33} p + N_r I_{13} r - 2Bz_b e_1 I_{33} - P_2^1 I_{13} \tau_2)/\Delta \\ (M_q q - 2Bz_b e_2 + P_1^3 \tau_1)/I_{22} \\ (L_p I_{13} p - N_r I_{11} r - 2Bz_b e_1 I_{13} + P_2^1 I_{11} \tau_2)/\Delta \\ u \\ v \\ w \\ -(e_1 p + e_2 q + e_3 r)/2 \\ p/2 \\ q/2 \\ r/2 \end{pmatrix} \qquad (8.61)$$

and $g(v, \eta, t)$ represents the remaining terms.

As the functions τ_1, τ_2, τ_3 are homogeneous of degree one with respect to the dilation and continuous for $(v, \eta) \neq 0$, they are continuous at zero. Furthermore, one can easily verify that $f(v, \eta, t)$ defines a periodic, continuous, homogeneous transformation of degree zero with respect to the dilation. Also, the function $g(v, \eta, t)$ is continuous and defines a sum of homogeneous field of degree strictly positive with respect to the dilation.

To prove the stability of the closed-loop system, it is sufficient to show that the origin of the unperturbed system

$$\begin{pmatrix} \dot{v} \\ \dot{\eta} \end{pmatrix} = f(v, \eta, t) \qquad (8.62)$$

is locally asymptotically stable.

For this purpose, let us consider the following reduced system obtained by this unperturbed system by taking $q = q_d$, $p = p_d$, $w = w_d$ as new control variables and where we have removed the equation corresponding to e_0 as it is uniquely defined by e_1, e_2, e_3 since the Euler parameters satisfy the equation $e_0^2 + e_1^2 + e_2^2 + e_3^2 = 1$ and we assume without loss of generality that $e_0 > 0$. Then, we obtain the following resulting system

$$\dot{u} = (X_u u + 2(B - mg)e_2)/m_{11} + (-M_q q_d + 2Bz_b e_2)/P_1^3 m_{11} \quad (8.63)$$

$$\dot{v} = (Y_v v - 2(B - mg)e_1 + m_{33}p_d w_d)/m_{22}$$
$$\qquad + (L_p I_{33} p_d - N_r I_{13} r + 2Bz_b e_1 I_{33})/P_2^1 I_{13} m_{22} \qquad (8.64)$$

$$\dot{r} = (L_p p_d + 2Bz_b e_1)/I_{13} \qquad (8.65)$$

$$\dot{x} = u \qquad (8.66)$$

$$\dot{y} = v \qquad (8.67)$$

$$\dot{z} = w_d \qquad (8.68)$$

$$\dot{e}_1 = p_d/2 \tag{8.69}$$
$$\dot{e}_2 = q_d/2 \tag{8.70}$$
$$\dot{e}_3 = r/2 \tag{8.71}$$

The control inputs p_d, q_d and w_d are given by equations (8.56)–(8.58). One can verify by application of Theorem 3.1 of [29] that the origin of the closed-loop system is asymptotically stable. Indeed, the vector field associated with right-hand side of the closed-loop system is continuous periodic and homogeneous of degree zero with respect to the dilation. Due to the periodic time-variant control, the resulting system is a periodic time-varying system, which can be written in the form

$$\begin{pmatrix} \dot{v} \\ \dot{\eta} \end{pmatrix} = h(v, \eta, t/\varepsilon) \tag{8.72}$$

We approximate this system by an averaged system which is autonomous. The averaged system is defined as

$$\begin{pmatrix} \dot{v} \\ \dot{\eta} \end{pmatrix} = h_0(v, \eta) \tag{8.73}$$

where

$$h_0(v, \eta) = \frac{1}{T'} \int_0^{T'} h(v, \eta, t/\varepsilon) \tag{8.74}$$

and where T' is the period. Now the corresponding averaged system is given by

$$\dot{u} = (X_u u + 2(B - mg)e_2)/m_{11}$$
$$+ (-M_q(-k^{e_2}e_2 - k^x x - k^u u) + 2Bz_b e_2)/P_1^3 m_{11} \tag{8.75}$$
$$\dot{v} = (Y_v v - 2(B - mg)e_1 - m_{33}k^z z(-k^{e_1}e_1 - k^{e_3}e_3 - k^r r))/m_{22}$$
$$+ (L_p I_{33}(-k^{e_1}e_1 - k^{e_3}e_3 - k^r r) - N_r I_{13}r + 2Bz_b e_1 I_{33})/P_2^1 I_{13} m_{22} \tag{8.76}$$
$$\dot{r} = (L_p(-k^{e_1}e_1 - k^{e_3}e_3 - k^r r) + 2Bz_b e_1)/I_{13} \tag{8.77}$$
$$\dot{x} = u \tag{8.78}$$
$$\dot{y} = v \tag{8.79}$$
$$\dot{z} = -k^z z \tag{8.80}$$
$$\dot{e}_1 = (-k^{e_1}e_1 - k^{e_3}e_3 - k^r r)/2 \tag{8.81}$$
$$\dot{e}_2 = (-k^{e_2}e_2 - k^x x - k^u u)/2 \tag{8.82}$$
$$\dot{e}_3 = r/2 \tag{8.83}$$

The linearization of the system around the origin is

$$\dot{u} = (X_u u + 2(B - mg)e_2)/m_{11} + (-M_q(-k^{e_2}e_2 - k^x x - k^u u)$$

$$+ 2Bz_b e_2)/P_1^3 m_{11} \tag{8.84}$$

$$\dot{v} = (Y_v v - 2(B - mg)e_1)/m_{22} - (-L_p I_{33}(-k^{e_1} e_1 - k^{e_3} e_3 - k^r r)$$
$$+ N_r I_{13} r - 2Bz_b e_1 I_{33})/P_2^1 I_{13} m_{22} \tag{8.85}$$

$$\dot{r} = (L_p(-k^{e_1} e_1 - k^{e_3} e_3 - k^r r) + 2Bz_b e_1)/I_{13} \tag{8.86}$$

$$\dot{x} = u \tag{8.87}$$

$$\dot{y} = v \tag{8.88}$$

$$\dot{z} = -k^z z \tag{8.89}$$

$$\dot{e}_1 = (-k^{e_1} e_1 - k^{e_3} e_3 - k^r r)/2 \tag{8.90}$$

$$\dot{e}_2 = (-k^{e_2} e_2 - k^x x - k^u u)/2 \tag{8.91}$$

$$\dot{e}_3 = r/2 \tag{8.92}$$

The above system can be reduced to the following subsystems

$$\dot{u} = (X_u u + 2(B - mg)e_2)/m_{11}$$
$$+ (-M_q(-k^{e_2} e_2 - k^x x - k^u u) + 2Bz_b e_2)/P_1^3 m_{11} \tag{8.93}$$

$$\dot{e}_2 = (-k^{e_2} e_2 - k^x x - k^u u)/2 \tag{8.94}$$

$$\dot{x} = u \tag{8.95}$$

and

$$\dot{v} = (-Y_v v - 2(B - mg)e_1)/m_{22} - (-L_p I_{33}(-k^{e_1} e_1 - k^{e_3} e_3 - k^r r)$$
$$+ N_r I_{13} r - 2Bz_b e_1 I_{33})/P_2^1 I_{13} m_{22} \tag{8.96}$$

$$\dot{r} = (L_p(-k^{e_1} e_1 - k^{e_3} e_3 - k^r r) + 2Bz_b e_1)/I_{13} \tag{8.97}$$

$$\dot{e}_1 = (-k^{e_1} e_1 - k^{e_3} e_3 - k^r r)/2 \tag{8.98}$$

$$\dot{e}_3 = r/2 \tag{8.99}$$

$$\dot{y} = v \tag{8.100}$$

Now, it is clear that for suitable gain parameters, the origin of each subsystem is obviously asymptotically stable. Therefore, the origin of the original system is locally asymptotically stable. Consequently, the origin of the system is asymptotically stable. The asymptotic stability of the origin of the system follows by direct application of Corollary 1 in [109], after noticing that the functions p_d, q_d and w_d are homogeneous of degree one with respect to the dilation and of class $\mathcal{C}^1$ on $\{\mathbb{R}^6 \times \mathbb{R}^3 - (0,0)\} \times \mathbb{R}$. $\qquad\square$

8.4 Simulation Results

The lighter-than-air platform is the AS200 by Airspeed Airships. It is a non-rigid 6 m long, 1.4 m diameter and 8.6 m^3 volume airship equipped with two vectorable engines on the sides of the gondola and four control surfaces at the stern. The four stabilizers are externally braced on the hull and rudder movement is provided by direct linkage to the servos. Envelope pressure is maintained by air fed from the propellers into the two ballonets located inside the central portion of the hull. These ballonets are self-regulating and can be fed from either engine. The engines are standard model aircraft type units. The propellers can be rotated through 120 degrees. During flight the rudder-vators (rudder and elevator) are used for all movements in pitch and yaw. In addition, the trim function can be used to alter the attitude of the airship in order to obtain level flight or to fly with a positive or negative pitch angle.

Guided by linear control theory applied to the linear approximation, we have chosen the following control parameters:

$$
\begin{aligned}
&k^u = 0.7 \quad &&k^v = 1.5 \quad &&k^w = 0.2 \\
&k^p = 0.35 \quad &&k^q = 0.01 \quad &&k^r = 2 \\
&k^x = 1 \quad &&k^y = 1.48 \quad &&k^z = 0.2 \\
&k^{e_1} = 0.1 \quad &&k^{e_2} = 1 \quad &&k^{e_3} = 1.1 \\
&\epsilon = 0.0002
\end{aligned}
$$

The initial position and orientation of the airship are taken as: $x_0 = 4$ $y_0 = 5$; $z_0 = 0$; $e_{1_0} = 0.1$; $e_{2_0} = 0.1$; $e_{3_0} = 0.1$. Initially the airship was at rest.

Figures 8.3 - 8.5 present the states of the system respectively along the x, y, and z axes. The velocities with respect to the inertial frame (global) are given in Figure 8.6.

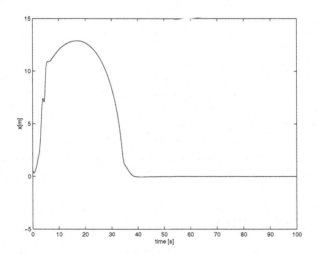

Fig. 8.3. Trajectory along the x-axis.

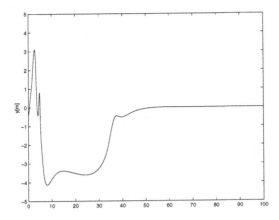

Fig. 8.4. Trajectory along the y-axis.

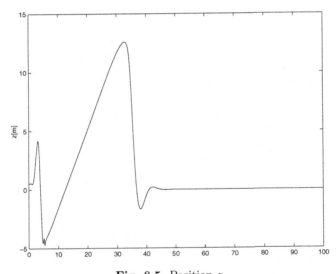

Fig. 8.5. Position z.

The simulations show that the airship converges to the origin. We notice that the gravitational force and moment are important for the stability properties of the airship.

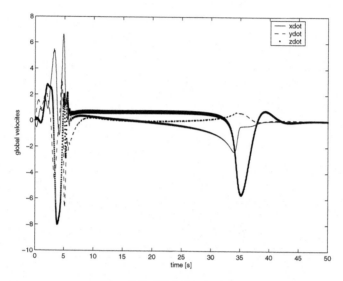

Fig. 8.6. Global velocities.

8.5 Conclusions

Airships represent a control challenge as they have non-zero drift. Their linearization at zero velocity is not controllable. We have proved that an airship represented by our model is not stabilizable by continuous state feedback. We have discussed the problem of stabilization of an airship and used the fact that the input vector fields are homogeneous of degree one with respect to some dilation. A feedback that is a homogeneous function of degree one makes the closed-loop vector field homogeneous of order zero. The main contribution of this work is an explicit smooth time-varying continuous feedback by using a time averaging technique. This feedback being uniformly stabilizing in time then each state may be bounded by a decaying exponential envelope. In this chapter, we have studied only local properties. In our future work, we will use the fact that asymptotic stability for the averaged system implies semi-global asymptotic stability for the actual system.

However, proper modelling of the other aerodynamic effects must be adopted. This airship can perform stabilization (station keeping) as well as short-range reconfiguration (tracking).

8.6 Nomenclature

A	homogeneous transform matrix
B	buoyancy
BG	position of the centre of buoyancy with respect to the body fixed frame
C	curve in $\mathbb{R}^3$
d	distance vector
D_V	diagonal aerodynamic force matrix
D_ω	diagonal aerodynamic moment matrix
e_3	unit vector along z
f	control force
g	constant gravity acceleration
I	identity matrix
J	inertia matrix
m	total mass of the airship
M	mass matrix
R	rotation matrix
s	curvilinear abscissa
$se(3)$	Lie algebra
$SE(3)$	Euclidean group
t	time
T	tangent unit vector
$V = (u\ v\ w)^T$	linear velocity vector
V_e	trim velocity
ρ_{air}	density of the air
ρ_{helium}	density of the helium
$\omega = (p\ q\ r)^T$	angular velocity vector
τ	control torque
Δ	volume of the envelope

9

Sensors, Modems and Microcontrollers for UAVs

The objective of this chapter is to present a brief description of the sensors currently used for measuring the position and orientation of an aerial vehicle.

Recently, a large variety of systems capable of measuring in 3D have been developed. A crucial point for mobile systems is the quality of the measures in a global reference mark. This quality is bound directly to the precision of the calibration and to the system of positioning used [42].

The positioning module is a vital component of any vehicle location and navigation system. To either help users obtain vehicle location or provide users with proper manoeuvre instructions, the vehicle location must be determined precisely. Therefore, accurate and reliable vehicle positioning is an essential prerequisite for any aerial vehicle stabilization [190].

Positioning involves determination of the coordinates of a vehicle on the surface of the Earth. Location involves the placement of the vehicle relative to landmarks or other terrain features such as roads [190].

To measure the position and orientation of a vehicle we can use different kinds of sensors. In general, there exist two types of sensors: relative sensors and absolute sensors. A relative sensor is a device that can measure the change in distance, position, or heading based on a predetermined or previous measurement. Without knowing an initial position (or previous reference) or heading, this sensor cannot be used to determine absolute position or heading with respect to the Earth. Absolute heading and position sensors are very important in solving location and navigation problems. An absolute sensor can provide information on the position of the vehicle with respect to the Earth. The most commonly used technologies for providing absolute position information are the magnetic compass and the GPS (Global Positioning System).

9.1 Polhemus Electromagnetic Sensor

The 3SPACE FASTRAK sensor (Polhemus) accurately computes the position and orientation of a tiny receiver as it moves through space. This device virtually eliminates the problem of latency as it provides dynamic, real-time six degree-of-freedom measurement of position (X, Y, and Z Cartesian coordinates) and orientation (ψ, θ, ϕ), and it is the most accurate electromagnetic tracking system available [46].

This sensor is ideal for measuring range of motion or limb rotation in biomedical research. It is a fast, accurate, easy to use, and effective method of capturing motion data on any non-conductive object. The Polhemus system utilizes a single transmitter and can accept data from up to four receivers. The use of advanced digital signal processing (DSP) technology provides an update rate of 120 Hz (with a single receiver) and a remarkable 4 ms latency. The data are then transmitted over a high-speed RS-232 interface at up to 115.2 K baud.

The Polhemus system uses electro-magnetic fields to determine the position and orientation of a remote object. The technology is based on generating near-field, low-frequency magnetic field vectors from a single assembly of three concentric, stationary antennae called a transmitter, and detecting the field vectors with a single assembly of three concentric, remote sensing antennae called a receiver. The sensed signals are input to a mathematical algorithm that computes the receiver's position and orientation relative to the transmitter.

Large metallic objects, such as desks or cabinets, located near the transmitter or receivers may adversely affect the performance of the system. Many walls, floors, and ceilings also contain significant amounts of metal.

This is an ideal sensor for indoor applications. The principal problems are that it is very sensitive to metallic objects and it works well only in a radius smaller than 76 cm.

9.1.1 Components

The Polhemus system includes a System Electronics Unit (SEU), a power supply, one to four receivers and one transmitter.

SYSTEM ELECTRONICS UNIT: Contains the hardware and software necessary to generate and sense the magnetic fields, compute position and orientation, and interface with the host computer via an RS-232.

TRANSMITTER: The transmitter is a triad of electromagnetic coils, enclosed in a plastic shell, that emits the magnetic fields. The transmitter is the system's reference frame for receiver measurements.

RECEIVER: The receiver is a small triad of electromagnetic coils, enclosed in a plastic shell, that detects the magnetic fields emitted by the transmitter. The receiver is a lightweight cube whose position and orientation are precisely measured as it is moved. The receiver is completely passive and highly reliable.

9.2 Inertial Navigation System

Every object that is free to move in space has six degrees of freedom. There are three degrees of freedom that specify its position (x, y, z) and three rotational degrees of freedom (ψ, θ, ϕ) that specify its attitude. If you know these six variables, you know where the object is and which way it is pointed. If you know them over a period of time, then you can also figure out how fast the object is moving, and what its acceleration rate is.

The inertial part of the Inertial Navigation System (INS)[1] is the way we obtain these variables. The operation of the inertial navigation system depends on the laws of classical mechanics formulated by Newton.

To built an INS, we essentially go backwards; we use accelerometers (devices that measure acceleration) and rate gyroscopes (devices which measure rotational velocity) to sense how the object is accelerating and rotating in space. We then get back to the position and attitude by integrating the accelerometers twice and the rate gyros once. We can use microcontrollers to do this integration. Using these sensors and the microcontroller, we can obtain the position and attitude of the aircraft. Then, we can use control theory to control the aircraft with thrusters or fins or whatever control mechanism we choose. That's where the navigation system part comes in. We're essentially closing the loop between the aircraft, the outside world, and where we want to be.

Generally an INS is composed of one or more Inertial Measurement Units (IMU) and Global Positioning Systems (GPS) to measure the position and the orientation of the drone. Some use the combination of other sensors like altimeters, barometers, inclinometers or vision, etc. The design of an inertial navigation system is a very complex issue and requires profound skills in analogue circuit design, signal processing and algorithm programming.

Inertial navigation is based on techniques that were invented and developed after the Second World War. The first systems were built of mechanical gyros, which required very complicated technical and power consuming constructions being prone to failure. Later on solid state solutions were realized by using only discrete integrated electro-mechanical or electro-optical sensors. These solid state systems had no moving parts (therefore Strapdown Inertial

---

[1] INS is a system to control an aircraft using inertial forces [39].

Navigation System), but consisted of expensive laser-gyros and integrated sensor devices in MEMS technology (Micro Electro-Mechanical System).

Inertial navigation systems are used in civil and military aviation, cruise missiles, submarines and space technology. In these areas of operation, the entire system and all components have to be very precise and reliable. As a consequence, the costs of such systems are still very high ($> 100,000$ USD) and the size is not yet small enough to be used for mobile roboting, wearable computing, automotive or consumer electronics. But applications designed for these industry branches require a very small and inexpensive implementation of a strapdown INS. Industrial demand for low-cost sensors (car airbag systems) and recent progress in MEMS integration technology led to sophisticated sensor products, which are now both small (single chip solution) and inexpensive (~ 100 USD).

The range of applications in which inertial navigation systems can and are being used is very extensive, covering navigation of ships, aircraft, tactical and strategic missiles and spacecraft. In addition, there are some more novel applications in the fields of robotics, active suspension in racing of high-performance motor cars and for surveying underground wells and pipelines [176].

Such diverse applications call for navigation systems with a very broad range of performance capabilities, as well as large differences in the periods of time over which they will be required to provide navigation data. Some applications may require inertial navigation and guidance to an accuracy of a few hundred metres or centimetres for periods of minutes or even a few seconds, while in other cases, the system may be required to provide navigation data to similar accuracy over periods of weeks, months or even longer.

Although the basic principles of inertial navigation systems do not change from one application to another, the instrument technologies and the techniques used for the implementation of the navigation function in such diverse applications vary greatly.

9.3 Accelerometers

DEFINITION

Accelerometers are sensors and instruments for measuring, displaying and analysing acceleration and vibration. They can be used on a stand-alone basis, or in conjunction with a data acquisition system. Accelerometers are available in many forms. They can be raw sensing elements, packaged transducers, or a sensor system or instrument, incorporating features such as totalizing, local or remote display and data recording.

Accelerometers can have from one to three axes of measurement, the multiple axes typically being orthogonal to each other [6]. These devices work on many operating principles. The most common types of accelerometers are piezoelectric, capacitance, null-balance, strain gauge, resonance, piezoresistive and magnetic induction.

9.3.1 Accelerometer Principles

There are several physical processes that can be used to develop a sensor to measure acceleration. In applications that involve flight, such as aircraft and satellites, accelerometers are based on properties of rotating masses. In the industrial world, however, the most common design is based on a combination of Newton's law of mass acceleration and Hooke's law of spring action [30].

SPRING-MASS SYSTEM

Newton's law simply states that if a mass, m, is undergoing an acceleration, a, then there must be a force F acting on the mass and given by $F = ma$. Hooke's law states that if a spring of spring constant k is stretched (extended) from its equilibrium position for a distance Δx, then there must be a force acting on the spring given by $F = k\Delta x$.

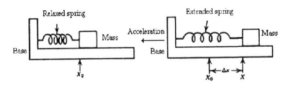

a) Spring-mass system with no acceleration   b) Spring-mass system with acceleration

Fig. 9.1. The basic spring-mass system accelerometer. Credits – National Instruments [30, 118].

In Figure 9.1 (a), we have a mass that is free to slide on a base. The mass is connected to the base by a spring that is in its unextended state and exerts no force on the mass. In Figure 9.1 (b), the whole assembly is accelerated to the left, as shown. Now the spring extends in order to provide the force necessary to accelerate the mass. This condition is described by equating Newton's and Hooke's laws:

$$ma = k\Delta x \tag{9.1}$$

where k is a spring constant in N/m, Δx is a spring extension in m, m is the mass in kg and a is the acceleration in m/s^2.

Equation (9.1) allows the measurement of acceleration to be reduced to a measurement of spring extension (linear displacement) because

$$a = \frac{k}{m} \Delta x \qquad (9.2)$$

If the acceleration is reversed, the same physical argument would apply, except that the spring is compressed instead of extended. Equation (9.2) still describes the relationship between spring displacement and acceleration.

The spring-mass principle applies to many common accelerometer designs. The mass that converts the acceleration to spring displacement is referred to as the test mass or seismic mass. We see, then, that acceleration measurement reduces to linear displacement measurement; most designs differ in how this displacement measurement is made.

NATURAL FREQUENCY AND DAMPING

On closer examination of the simple principle just described, we find another characteristic of spring-mass systems that complicates the analysis. In particular, a system consisting of a spring and attached mass always exhibits oscillations at some characteristic natural frequency. Experience tells us that if we pull a mass back and then release it (in the absence of acceleration), it will be pulled back by the spring, overshoot the equilibrium, and oscillate back and forth. Only friction associated with the mass and base eventually brings the mass to rest. Any displacement measuring system will respond to this oscillation as if an actual acceleration occurs. This natural frequency is given by

$$f_N = \frac{1}{2\pi} \sqrt{\frac{k}{m}} \qquad (9.3)$$

where f_N is the natural frequency in Hz, k is a spring constant in N/m and m is the seismic mass in kg.

The friction that eventually brings the mass to rest is defined by a damping coefficient, which has the units s^{-1}. In general, the effect of oscillation is called the transient response, described by a periodic damped signal, as shown in Figure 9.2, whose equation is

$$X_T(t) = X_o e^{-\mu t} \sin(2p f_N t) \qquad (9.4)$$

where $X_T(t)$ is the transient mass position, X_o is the peak position, initially, μ is the damping coefficient, f_N is the natural frequency and t is the time.

The parameters, natural frequency, and damping coefficient in equation (9.4) have a profound effect on the application of accelerometers.

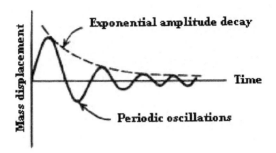

Fig. 9.2. A spring-mass system exhibits a natural oscillation with damping as response to an impulse input. Credits – National Instruments [30, 118].

VIBRATION EFFECTS

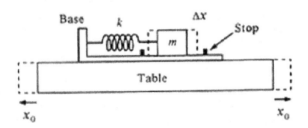

Fig. 9.3. A spring-mass accelerometer has been attached to a table which is exhibiting vibration. The table peak motion is x_o and the mass motion is Δx. Credits – National Instruments [30, 118].

The effect of natural frequency and damping on the behaviour of spring-mass accelerometers is best described in terms of an applied vibration. If the spring-mass system is exposed to a vibration, then the resultant acceleration of the base is given by

$$a(t) = -\omega^2 x_o \sin(\omega t) \tag{9.5}$$

where $a(t)$ is the vibrating acceleration, x_o is the peak displacement from equilibrium in m and w is the angular frequency in rad/s.

If this is used in equation (9.1), we can show that the mass motion is given by

$$\Delta x = \frac{-mx_o}{k} \omega^2 \sin(\omega t) \qquad (9.6)$$

where $\omega = 2\pi f$.

To make the predictions of equation (9.6) clear, consider the situation presented in Figure 9.3. Our model spring-mass accelerometer has been fixed to a table that is vibrating.

The x_o in equation (9.6) is the peak amplitude of the table vibration, and Δx is the vibration of the seismic mass within the accelerometer. Thus, equation (9.6) predicts that the seismic-mass vibration peak amplitude varies as the vibration frequency squared, but linearly with the table-vibration amplitude. However, this result was obtained without consideration of the spring-mass system's natural vibration. When this is taken into account, something quite different occurs.

Figure 9.4 (a) shows the actual seismic-mass vibration peak amplitude versus table-vibration frequency compared with the simple frequency squared prediction. You can see that there is a resonance effect when the table frequency equals the natural frequency of the accelerometer, that is, the value of Δx goes through a peak. The amplitude of the resonant peak is determined by the amount of damping. The seismic-mass vibration is described by equation (9.6) only up to about $f_N/2.5$.

Figure 9.4 (b) shows two effects. The first is that the actual seismic-mass motion is limited by the physical size of the accelerometer. It will hit stops built into the assembly that limit its motion during resonance. The figure also shows that for frequencies well above the natural frequency, the motion of the mass is proportional to the table peak motion, x_o, but not to the frequency.

Thus, it has become a displacement sensor. To summarize:

1. $f < f_N$ For an applied frequency less than the natural frequency, the natural frequency has little effect on the basic spring-mass response given by equations (9.1) and (9.6). A rule of thumb states that a safe maximum applied frequency is $f < \frac{1}{2.5f_N}$.

2. $f > f_N$ For an applied frequency much larger than the natural frequency, the accelerometer output is independent of the applied frequency. As shown in Figure 9.4 (b), the accelerometer becomes a measure of vibration displacement x_o of the following equation

$$x(t) = x_o \sin(\omega t) \qquad (9.7)$$

where $x(t)$ is the object position in m.

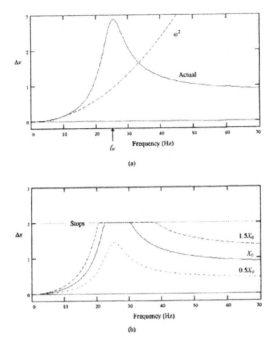

Fig. 9.4. In (a) the actual response of a spring-mass system to vibration is compared with the simple w^2 prediction. In (b) the effect of various table peak motion is shown. Credits – National Instruments [30, 118].

It is interesting to note that the seismic mass is stationary in space in this case, and the housing, which is driven by the vibration, moves about the mass. A general rule sets $f > 2.5 f_N$ for this case.

Types of accelerometers

In terms of functionality, accelerometers can be divided into those that require an external power supply and those that do not, and into those that respond to static accelerations (such as the acceleration due to gravity) and those that do not. We describe the most used.

POTENTIOMETRIC ACCELEROMETER

This simplest accelerometer type measures mass motion by attaching the spring mass to the wiper arm of a potentiometer (Figure 9.5). In this manner, the mass position is conveyed as a changing resistance. The natural frequency of these devices is generally less than 30 Hz, limiting their application to steady-state acceleration or low-frequency vibration measurement. Numerous signal-conditioning schemes are employed to convert the resistance variation into a voltage or current signal.

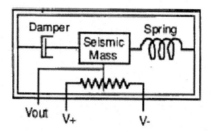

Fig. 9.5. Potentiometric accelerometer.

SERVO OR FORCE BALANCE ACCELEROMETER

These types of accelerometers are typically used where accuracy is essential in the low g ranges such as in dead-reckoning navigation systems. *Servo* or *force balance* indicates the design philosophy used to create the accelerometer rather than any particular technology or fabrication method; in such a device, the displacement of the seismic mass is detected by a displacement sensor (Figure 9.6). The signal from the displacement sensor is used to restore the mass to its original position as well as to obtain a direct measure of the applied acceleration.

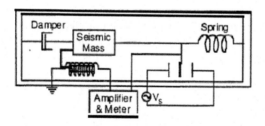

Fig. 9.6. Force balance accelerometer.

INDUCTIVE ACCELEROMETER

A mass is suspended between two iron core inductance coils (Figure 9.7). The induction coils are part of an alternating current Wheatstone bridge so that as the mass is displaced due to acceleration, a voltage change proportional to the magnitude of the acceleration is produced. These devices are character-ized by good long-term stability, but they cannot be easily produced in large quantities.

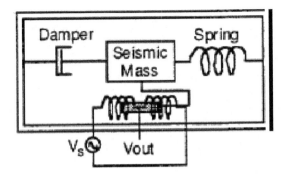

Fig. 9.7. Inductive accelerometer.

ELECTROLYTIC ACCELEROMETER

An electrolytic liquid is restrained in a curved tube much like that of a common level (Figure 9.8). Two metal probes, inserted from each end, pass through the liquid and into the bubble. A third probe is fully submerged in the liquid. Changes in acceleration are measured by the movement of the bubble. As long as no net acceleration is acting on the bubble, it remains stationary. If the bubble shifts in one direction or the other due to acceleration, a corresponding difference in resistance is measured. In comparison to the classical accelerometer, the damping function is provided by the viscous liquid, the spring function by a combination of gravity and buoyancy, and the seismic mass by the liquid (not the air bubble). These are simple devices used to measure small slowly varying horizontal accelerations and do not lend themselves to an automated collision notification (ACN) system.

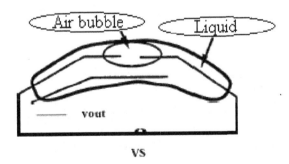

Fig. 9.8. Electrolytic accelerometer.

CAPACITIVE ACCELEROMETER

This device has a parallel plate capacitor arrangement in which one plate of the capacitor is free to move with applied stress (Figure 9.9). During an acceleration, the distance between the movable plate and the fixed plates changes, thus changing the capacitance between the movable and fixed plates. The change in capacitance is directly related to the acceleration. Capacitive accelerometers can be produced using surface micromachining techniques that allow the capacitor and all signal conditioning electronics to be fabricated on a single silicon chip using standard integrated circuit manufacturing techniques.

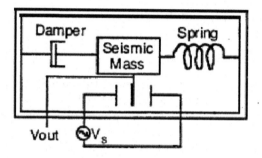

Fig. 9.9. Capacitive accelerometer.

STRAIN-GAUGE ACCELEROMETER

These devices are composed of a mass and metallic spring element (Figure 9.10). The spring element utilizes conventional strain gauges, usually in a Wheatstone bridge configuration. When acceleration acts on the mass and deforms the spring, the resulting strain gauge output is proportional to the applied acceleration. Although common and relatively inexpensive, this type of sensor cannot be easily produced in large quantities.

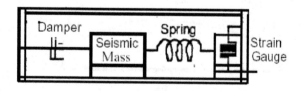

Fig. 9.10. Strain-gauge accelerometer.

PIEZORESISTIVE ACCELEROMETERS (VARIABLE CAPACITANCE ACCELEROM-
ETERS)

Piezoresistive sensors are typically made from a surface micromachined polysilicon structure on which polysilicon springs, whose electrical resistance changes
with acceleration forces, are arranged in a Wheatstone bridge configuration
(Figure 9.11). An applied acceleration produces a voltage proportional to the
amplitude of the acceleration. Variable capacitance accelerometers typically
use a differential capacitor with central platter attached to the moving mass
and fixed external plates. An applied acceleration unbalances the capacitor,
resulting in an output wave with amplitude proportional to the applied acceleration. Both of these types of accelerometers use MEMS technology, which
results in them being miniature, low-cost instruments. Both types require an
external power supply that allows the accelerometer to respond to static accelerations [101]. There are two primary disadvantages with these devices.
First, they must be bulk micromachined, which does not allow the supporting electronics to be placed on the same chip. Second, they are sensitive to
temperature [117].

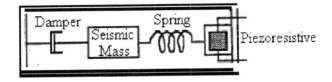

Fig. 9.11. Piezoresistive accelerometer.

PIEZOELECTRIC ACCELEROMETERS

The piezoelectric accelerometer is based on a property exhibited by certain
crystals where a voltage is generated across the crystal when stressed. This
property is also the basis for such familiar sensors as crystal phonograph
cartridges and crystal microphones.

Mounted within the accelerometer housing is a piezoelectric crystal af-
fixed to a small mass (Figure 9.12). This mass is coupled to the supporting
accelerometer base through the piezoelectric crystal. When the accelerometer
is subjected to acceleration, the mass exerts a force on the crystal. This force
results in a charge output from the crystal that is directly proportional to the
input acceleration.

There are many piezoelectric crystal materials that are useful in accelerometer construction. The most common materials used are ceramic lead metaniobate, lead zirconate, lead titanate and natural quartz crystal. There are also many different mechanical configurations of the masses and crystals within the accelerometer case. Typical configurations include isolated compression, shear, and ring shear.

The materials and physical construction used in an accelerometer design are selected to furnish the particular performance characteristics desired.

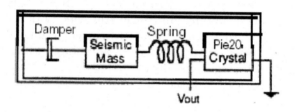

Fig. 9.12. Piezoelectric accelerometer.

9.3.2 Applicability of Accelerometers

Accelerometers are recommended for different systems including aircraft control. This is not only due to their ability to completely characterize the forces but also their increased reliability, testability, reduced size, and reduced cost. Of the eight accelerometer types, the micromachined solid state devices such as the capacitive and piezoresistive accelerometers are the most acceptable candidates from both cost and performance perspectives.

The most common applications are:

- Inertial measurement of velocity and position
    - Acceleration single integrated for velocity
    - Acceleration double integrated for position
- Vibration and shock measurement
    - Measuring vibration for machine health
    - Motion and shock detection
- Measurement of gravity to determine orientation
    - Tilt and inclination
    - Position in 2– and 3–dimensional space

9.4 Inclinometers

Definition

1. An inclinometer is a gravity reference device capable of sensing tilt. The instrument is basically a special implementation of a linear accelerometer with low maximum acceleration capability. The accelerometer output is usually processed to give a d.c. voltage directly proportional to the angle of tilt. Typical applications of the inclinometer are platform levelling for target acquisition systems and fire control systems, and in inertial component testing.

2. The inclinometer is a unidirectional sensor using a capacitive transducer. The output is a variable square wave which is proportional to the angle. The dielectric used within the transducer is chosen to provide damping by reducing oscillations caused by overshooting or shock, yet still providing a reasonable reaction time to angular movements.

9.5 Altimeters

Definition

An altimeter is an active instrument used to measure the altitude of an object above a fixed level (Figure 9.13). For example, a laser altimeter can measure height from a spacecraft above an ice-sheet. That measurement, coupled with radial orbit knowledge, will enable determination of the topography [185].

The traditional altimeter found in most aircraft works by measuring the air pressure from a static port in the airplane. Air pressure falls almost linearly with altitude about one millibar per 30 feet at typical flying levels. The altimeter is calibrated to show the pressure directly as altitude. The pilot must adjust the altimeter to account for the local air pressure at ground level, which varies with the weather.

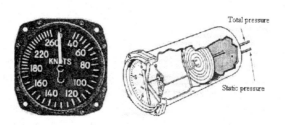

Fig. 9.13. The altimeter.

In pilot's jargon, the regional air pressure at mean sea level is called the QNH,[2][3] and the pressure which will calibrate the altimeter to show the height above ground at a given airfield is called the QFE[4] of the field.

Another type of altimeter is the radar altimeter which measures the altitude more exactly using the time taken for a radio signal to reflect from the surface back to the aircraft. The radar altimeter is used to measure the exact height during the landing procedure of commercial aircraft.

Barometric altimeters are invariably used for height measurement in aircraft. As supplementary navigation aids, they are widely used for restricting the growth of errors in the vertical channel of an inertial navigation system. In a Schuler tuned inertial navigation system, although the propagation of errors in the horizontal channels is bounded, the velocity and position errors in the vertical channel are not bounded and can become very large within a relatively short period of time unless there is an independent means of checking the growth of such errors [176].

---

[2] QNH is a Q code used by pilots, air traffic control (ATC) and low-frequency weather beacons to refer to the current mean air pressure over a given region at sea level (if there is no sea, this is a virtual value by adjusting the value at the ground for its elevation), the "regional pressure setting". This value is used by pilots to calibrate the altimeter on board the aircraft, to ensure that the pilot is accurately aware of his actual flying height. The QNH allows a pilot to know his altitude with respect to mean sea level [185].

[3] The Q code is a set of three-letter code signals to be used in radiotelegraphy and amateur radio communications. It was developed and instituted in 1912 as a way to facilitate communication between maritime radio operators of different nationalities. For this reason, callsigns never begin with a Q [185].

[4] QFE is a Q code used by pilots and air traffic control to refer to the current air pressure. When dialled into the pilot's altimeter the instrument will read the height above the ground at a particular airfield. This setting is used during take-off and landing and when flying in the circuit. A mnemonic for the code is *Q Field Elevation* [185].

A barometric altimeter, relying on atmospheric pressure readings, provides an indirect measure of height above a nominal sea level, typically to an accuracy of 0.1%. Most airborne inertial systems requiring a three-dimensional navigation capability operate with barometric aiding in order to bound the growth of vertical channel errors [176].

A radio altimeter provides a direct measure of height above ground which is equally important for many applications. Such measurement may be used in conjunction with a stored map of the terrain over which an aircraft is flying to provide position updates for an inertial navigation system [176].

9.6 Gyroscopes

Fig. 9.14. A gyroscope.

DEFINITION

1. A rotating mechanism in the form of a universally mounted spinning wheel that offers resistance to turns in any direction.

2. A rotating wheel, mounted in a ring or rings, for illustrating the dynamics of rotating bodies, the composition of rotations, etc. It was devised by Professor W. R. Johnson, in 1832, by whom it was called the *rotascope*.

3. A form of the above apparatus, invented by M. Foucault, mounted so delicately as to render visible the rotation of the earth, through the tendency of the rotating wheel to preserve a constant plane of rotation, independently of the earth's motion.

4. A device which utilizes the angular momentum of a spinning mass (rotor) to sense angular motion of its base about one or two axes orthogonal to the spin axis. Also called a gyro.

A gyroscope (Figure 9.14) exhibits a number of behaviours including precession.[5] Gyroscopes can be used to construct gyrocompasses which replace magnetic compasses (in aircraft and spacecraft), to assist in stability (bicycle, Hubble Space Telescope), as a repository for angular momentum (momentum wheels) and as a means to conserve and deliver energy (flywheel energy storage) in some machines. The flywheel in a engine is one such use. Gyroscopic effects are used in many different toys, such as yo-yos and dynabees.

Gyroscopes can be very perplexing objects because they move in peculiar ways and even seem to defy gravity. These special properties make gyroscopes extremely important in everything from your bicycle to the advanced navigation system on the space shuttle. A typical airplane uses about a dozen gyroscopes in everything from its compass to its autopilot. The Russian Mir space station used 11 gyroscopes to keep its orientation to the sun, and the Hubble Space Telescope has a batch of navigational gyros as well.

Gyroscopes are used in various applications to sense either the angle turned through by a vehicle or structure (displacement gyroscopes) or, more commonly, its angular rate of turn about some defined axis (rate gyroscopes) [58, 142]. The sensors are used in a variety of roles such as:

*   Flight path stabilization
*   Autopilot feedback
*   Sensor or platform stabilization
*   Navigation.

Examples of mechanical spinning wheel gyroscopes used in strapdown applications are the single-axis rate integrating gyroscope and twin-axis tuned or flex gyroscopes. An alternative class designation for gyroscopes that cannot be categorized in this way is "unconventional sensors", some of which are solid state devices. The very broad and expanding class of unconventional sensors includes devices such as:

*   Rate transducers which include mercury sphere and magneto-hydrodynamic sensors
*   Vibratory gyroscopes
*   Nuclear magnetic resonance (NMR) gyroscopes
*   Electrostatic gyroscopes (ESGs)
*   Optical rate sensors which include ring laser gyroscopes (RLGs) and fibre optic gyroscopes (FOGs).

---

[5] Precession is the phenomenon by which the axis of a spinning object wobbles when a torque is applied to it. The phenomenon is commonly seen in a spinning toy top, but all rotating objects can undergo precession [39].

Although many of the sensors in this class are strictly angular rate sensors and not gyroscopes in the sense that they do not rely on the dynamical properties of rotating bodies, it has become accepted that all such devices are referred to as gyroscopes since they all provide measurements of body rotation.

9.6.1 Types of Gyroscopes

Conventional sensors

SPINNING MASS GYRO

This is the classical gyro, which has a mass spinning steadily with a free movable axis (called a gimbal). When the gyro is tilted, the gyroscopic effect causes precession (motion orthogonal to the direction of tilt) on the rotating mass axis, hence letting you know the angle moved. Because mechanical constraints cause numerous error factors, by fixing the axis with springs, the spring tension is proportional to the precession speed. By integrating the spring tension one gets the angle. An angular velocity (rate of turn) sensor, therefore, is a rate-gyroscope. Nowadays most gyroscopes are actually rate-gyroscopes. A dry tuned gyro (dynamically tuned gyro) is a type of spinning mass gyro, which has been designed to cause very small mechanical constraints once the spinning speed reaches a specific speed. This should not be confused with a gyrocompass [72]. A gyrocompass is also a spinning mass gyro (usually big), but its axis is made to rotate in (and maintain) the same direction as that of the earth rotation all the time, hence giving True North at all times, but a gyroscope (gyro) gives you information on the relative change of angles. The sale of spinning mass gyros is always contingent on specific regular maintenance.

The performance that may be achieved using gyroscopes of this type varies from precision devices with error rates of less than $0.001°$/hour, to less accurate sensors with error rates of tens of degrees per hour. Many devices of this type have been developed for strapdown applications, being able to measure angular rate up to about $500°$/second. Some designs are very rugged, having characteristics which allow them to operate in harsh environments such as guided weapons.

All gyroscope sensors are subject to errors which limit the accuracy to which the angle of rotation or applied turn rate can be measured. Spurious and undesired torques, caused by design limitations and constructional deficiencies, act on the rotors of all mechanical gyroscopes. These imperfections given rise to precession of the rotor, which manifests itself as a drift in the reference direction defined by the spin axis of the rotor. In a free gyroscope, i.e. one which measures angular displacements from a given direction, it is customary to describe the performance in terms of an angular drift rate. For a restrained gyroscope, i.e. one operating in a nulling or rebalance loop mode to provide a measure of angular rate, any unwanted torques act to produce a bias on the measurement of angular rate.

We use the term drift for the motion of the spin axis in a free gyroscope, and the term bias is used with nulled sensors. In practice, the way in which the errors are quoted often depends on the accuracy band of the sensor rather than whether the gyroscope is used with its spin axis fixed in space or restrained in some way.

The major sources of error which arise in mechanical gyroscopes are:

- Fixed bias (g-independent):

  The sensor output which is present even in the absence of an applied input rotation. It may be a consequence of a variety of effects, including residual torques from flexible leads within the sensor, spurious magnetic fields and temperature gradients.

- Acceleration-dependent bias (g-dependent):

  Biases which are proportional to the magnitude of the applied acceleration. Such errors arise in spinning mass gyroscopes as a result of mass imbalance in the rotor suspension.

- Anisoelastic bias (g^2-dependent):

  Biases which are proportional to the product of acceleration along orthogonal pairs of axes.

- Anisoinertia errors:

  Such errors arise in spinning mass gyroscopes and introduce biases owing to inequalities in a gyroscope's moments of inertia about different axes. Anisoinertia is frequency sensitive if the rotor is driven by a hysteresis motor.

- Scale factor errors:

  Errors in the ratio relating the change in the output signal to a change in the input rate which is to be measured.

- Cross-coupling errors:

  Erroneous gyroscope outputs resulting from gyroscope sensitivity to turn rates about axes normal to the input axis.

- Angular acceleration sensitivity:

  This error is also known as the gyroscopic inertial error. All mechanical gyroscopes are sensitive to angular acceleration owing to the inertia of the rotor. Such errors become important in wide bandwidth applications. This error increases with increasing frequency of input motion.

RATE INTEGRATING GYROSCOPE

A rate integrating gyroscope has one input axis and so is known as a single-axis gyroscope. The basic concept is capable of achieving a wide spectrum of performance from a very small gyroscope that fits into a cylinder of diameter 25 mm and length 50 mm. Typically, the drift performance of the miniature versions of this type of sensor is in the $1°$/hour to $10°$/hour class, although substantially better than $0.01°$/hour can be achieved with the larger top of the range sensors. The smaller sensors are able to measure turn rates typically of the order of $400°$/second or better. This type of sensor has found many different applications as a result of this wide spectrum of performance, including navigation systems in aircraft, ships and guided weapons.

DYNAMICALLY TUNED GYROSCOPE

This sensor is sometimes also called the tuned rotor gyroscope, or dry tuned gyroscope. It has two input axes which are mutually orthogonal and which lie in a plane which is perpendicular to the spin axis of the gyroscope. Generally, the performance of these gyroscopes is very similar to that achieved by rate integrating gyroscopes. Miniature instruments of this types developed from strapdown applications are typically about 30 mm in diameter and 50 mm long. Sub-miniature devices have also been produced, with some slight degradation in performance, which are about 15 mm by 35 mm. These gyroscopes have found many applications similar to the floated rate integrating gyroscope.

FLEX GYROSCOPE

This sensor bears a close resemblance to the dynamically tuned gyroscope and operates in a similar manner, as the rotor acts as a free inertial element. It also has two sensitive input axes. The form of construction allows a very small instrument to be made, typically about 20 mm in diameter and 30 m long. These sensors have found many applications in aerospace and industrial applications.

Rate sensors

There is a class of mechanical sensors designed to sense angular rate using various physical phenomena which are suitable for use in some strapdown applications. Such devices resemble conventional gyroscopes in that they make use of the principles of gyroscopic inertia and precession. They are suitable for some lower accuracy strapdown applications, particularly those that do not require navigational data, but require stabilization. These devices tend to be rugged and to be capable of measuring rotation rates up to about 500°/second with typical drift accuracies of a few hundred degrees per hour.

DUAL-AXIS RATE TRANSDUCER (DART)

This gyroscope has the ability to sense angular rate about two orthogonal axes. Its basic performance is certainly sub-inertial, typically having a drift in the region of 0.5°/second or less. Its size is somewhat smaller than the rate integrating gyroscope being about 18 mm in diameter and 40 mm long.

The inertial element in this form of gyroscope is a sphere of heavy liquid, such as mercury, contained in a spherical cavity. This cavity is rotated at high speed about an axis along the case in order to give high angular momentum to the fluid sphere. There is an assembly of paddles, rigidly mounted on the inside of this spherical cavity. These paddles have piezoelectric crystals attached to them. The instrument is sensitive to angular rates of the case about two orthogonal axes normal to the spin axis.

MAGNETO-HYDRODYNAMIC SENSOR

The sensor consists of an angular accelerometer and a synchronous motor. This sensor does not rely upon the angular momentum of a spinning mass, but uses a rotating angular accelerometer to sense angular rates about two mutually perpendicular axes of the sensor. The rotating angular accelerometer acts as an integrator and provides an electrical signal directly proportional to the applied angular rate. It is of similar size to the dual-axis rate transducer and has comparable performance capability, the g-independent bias being in the region of 0.05°/second to 0.5°/second.

Vibratory gyroscopes

The basic principle of operation of this sensor is that the vibratory motion of part of the instrument creates an oscillatory linear velocity. If the sensor is rotated about an axis orthogonal to this velocity, a Coriolis acceleration is induced. This acceleration modifies the motion of the vibrating element and provided that this can be detected, it will indicate the magnitude of the applied rotation.

The most common design technology for these sensors has generally used a stable quartz resonator with piezoelectric driver circuits. Some designs have produced sensors with small biases, in the region of 0.1 °/hour. Typical limitations for this type of technology, for use in inertial navigation systems, have been high drift rates, resonator time constants and sensitivity to environmental effects, particularly temperature changes and vibratory motion. However, these sensors can be made to be extremely rugged, including the capability of withstanding applied accelerations of many tens of thousands of g.

These sensors are usually quite small, usually with a diameter of something less than 15 mm and a length of about 25 mm. These sensors have been used in many applications, particularly to provide feedback for stabilization or angular position measurement tasks.

All vibrating sensors tend to have a very short reaction time, i.e. rapid start-up capability, and some designs are very rugged. Significant sources of error with these devices are their sensitivity to changes in ambient temperature and the potential for cross-talk between similar sensors mounted on the same structure. These devices are usually termed solid-state sensors and offer good shelf-life and good dormancy characteristics as they have no bearings, lubricants or any other fluid within their case. In addition, such sensors can be sensitive to vibration although, with careful design, such effects can be minimized.

The different types of design of such sensors are:

VIBRATING WINE GLASS SENSOR

This sensor is synonymous with vibrating cylinder and vibrating dome gyroscopes. The resonant body, usually a hemisphere or cylinder, is forced to vibrate at its resonant frequency by four equally spaced piezoelectric driving crystals firmly attached to its circumference.

VIBRATING DISC SENSOR

In this sensor, the resonator is formed from a metal alloy disc, which is machined to form a ring that is supported by rigid spokes. This ring is forced into resonant sinusoidal oscillation in the plane of the ring, using an alternating magnetic field, creating distortions in the shape of the ring. The motion of the ring is detected using capacitive techniques to measure the distance between a fixed plate and the edge of the ring.

TUNING FORK SENSOR

This form of device is very similar to the wine glass sensor. The sensing element is two vibrating structures mounted in parallel on a single base, each structure having a mass positioned at the end of a flexible beam. When the two structures are excited to vibrate in opposition, the effect is analogous to the motion of the tines of a tuning fork.

QUARTZ RATE SENSOR

The quartz rate sensor is a direct application of the tuning fork principle. It is a single degree-of-freedom, open-loop, solid-state sensor. In this device, quartz is formed into an H fork configuration, where one pair of tines has an array of electrodes. These tines are driven at their resonant frequency of about 10 kHz.

SILICON SENSOR

The material silicon has many properties that make it suitable for the fabrication of very small components and intricate monolithic devices. It is inexpensive, very elastic, non-magnetic, it has a high strength to weight ratio and possesses excellent electrical properties allowing component formation from diffusion or surface deposition. Additionally, it can be electrically or chemically etched to very precise tolerances, of the order of micrometres.

VIBRATING WIRE RATE SENSOR

This device, also known as the vibrating string gyroscope, has three fundamental components: a vibrating element in the form of a taut conductor, a drive magnet and a signal or pick-off magnet.

Cryogenic devices

NUCLEAR MAGNETIC RESONANCE GYROSCOPE

The nuclear magnetic resonance gyroscope has many attractions, particularly as it does not have any moving parts. Its performance is governed by the characteristics of the atomic material and does not demand the ultimate in accuracy from precision engineering techniques. Hence, in theory, it offers the prospect of a gyroscope with no limit to either its dynamic range or linearity and therefore, potentially, is an ideal sensor for use in strapdown inertial navigation. Nuclear magnetic resonance (NMR) is a physical effect arising from the interaction between the nuclei of certain elements and an external magnetic field.

SARDIN (SUPERCONDUCTING ABSOLUTE RATE DETERMINING INSTRUMENT)

The device is basically a superconducting cylindrical capacitor, the behaviour of which is governed by the principle that a closed superconducting ring always keeps the amount of flux linking it at a constant value. It does this by generating super-currents which flow in the ring for an indefinite length of time encountering no resistance to their motion.

Optical gyroscope

This term is applied to those classes of gyroscope which use the properties of electromagnetic radiation to sense rotation. Optical gyroscopes use an interferometer or interferometric methods to sense angular motion. In effect, it is possible to consider the electromagnetic radiation as the inertial element of these sensors.

The spectrum performance of optical gyroscopes ranges from the very accurate with a bias of less than 0.001 °/hour, usually ring laser, to tens of degrees per hour, often from simple fibre optic gyroscopes.

The types of optical gyroscopes are:

- Ring laser gyroscope
- Fibre optic gyroscope
- Fibre optic ring resonator gyroscope
- Ring resonator gyroscope.

Other gyroscopes

Other gyroscopes include gas rate gyros, etc. The gas rate gyro sprays gas onto heated wires. When there is rotation the spray is curved causing a change in the temperature of the wires. Numerous factors including fluid mechanics uncertainties prevent the practical usage of gas rate gyros, although they have been produced in several cases with enclosures bigger than that typical for vibrating gyros. Characteristics change with mounting position, due to convectional heat transfer.

9.6.2 Uses of Gyroscopes

There are a substantial number of gyro applications for which a drift rate of around 1 degree/second is not unacceptable, but low cost, low weight and volume, high maximum rate, high reliability and good dormancy are of paramount importance. Existing mechanical instruments of this class usually fail to meet all these requirements, with the maximum rate being a particular problem.

The effect of all this is that, once you spin a gyroscope, its axis wants to keep pointing in the same direction. If you mount the gyroscope in a set of gimbals so that it can continue pointing in the same direction, it will. This is the basis of the gyro-compass. If you mount two gyroscopes with their axes at right angles to one another on a platform, and place the platform inside a set of gimbals, the platform will remain completely rigid as the gimbals rotate in any way they please. This is this basis of inertial navigation systems (INS).

In an INS, sensors on the gimbals' axes detect when the platform rotates. The INS uses those signals to understand the vehicle's rotations relative to the platform. If you add to the platform a set of three sensitive accelerometers, you can tell exactly where the vehicle is heading and how its motion is changing in all three directions. With this information, an airplane's autopilot can keep the plane on course, and a rocket's guidance system can insert the rocket into the desired orbit.

The most common applications for gyroscopes are:

SCIENCE DEMONSTRATIONS

Demonstration gyroscopes are often found in learning environments such as schools or colleges to teach the physics of gyroscopes. Traditional demonstration gyroscopes tend to be gimballed to allow the user to understand how a gyroscope can continually point in one direction. A gimballed gyroscope allows the user to place weights/forces to one axis to see how the gyroscope will react. The user can also put pressure on a side of the gyroscope (by touching the gimbal) and "feel" the forces involved, which gives a little more excitement to the demonstration.

COMPUTER POINTING DEVICES

There are a number of computer pointing devices (in effect, a mouse) on the market that have gyroscopes inside them allowing you to control the mouse cursor while the device is in the air. They are also wireless so are perfect for presentations when the speaker is moving around the room. The gyroscope inside tracks the movements of your hand and translates them to cursor movements.

IN-CAR NAVIGATION SYSTEMS

Gyroscopic behaviour is used in the racing car industry. This is because car engines act just like big gyroscopes. This has its uses, for example in American Indy car racing some of the tracks are oval in shape. During the race the cars go round the circuit in one direction only (the car only ever turns in one direction, e.g. left). Because of the gyroscopic forces from the engine, depending on whether the engine is spinning clockwise or anti-clockwise, the car's nose will be forced up or down. Providing the engine spins in the right direction it can help the car to stay on the track.

GYROCOMPASSES

Gyrocompasses are basically navigation aids. Gyroscopes have a tendency to preserve their axis of rotation, so if they are mounted into a device that allows them to move freely (low-friction gimbal), when the device is moved in different directions the gyroscope will still point in the same direction. This can be measured and the results can be used in similar ways to a normal compass, but unlike a standard magnetic compass this is not affected by magnetic environmental changes and readings are more accurate. Gyrocompasses are commonly used in ships and aircraft.

ANTI-ROLL DEVICES/STABILIZERS

Again because of this behaviour, gyroscopes are used to stop things from falling over; some mono-trains use gyroscopes and ships often use them so that in bad seas the ship is kept relatively upright and not thrown about.

ARTIFICIAL HORIZONS/AUTOPILOTS

This works in the same way as a gyrocompass but on a different axis. The artificial horizon gauge shows the position of the aircraft relative to the horizon, i.e. the pitch of the aircraft.

Other possibles uses for gyroscopes are:

- Automotive yaw rate sensor
- Global Positioning System (GPS-INS, RTK) dead-reckoning
- Fluxgate compass compensation
- Mobile dish antenna stabilized platform
- Mobile camera platform stabilization
- Robotics
- Simulators
- Telematics / fleet-management
- Remote control vehicles
- Inertial / dynamic measurement units
- Drive safety recorder (blackbox for car/marine)
- AGV, wheelchair
- Marine satellite compasses
- Angle wrench
- Agricultural tractors
- Physiotherapy / therapeutic equipment
- 3-D input devices.

Type	Performance	Cost	Size
Mechanical	Very good	Expensive	Large
Optical	Very good	Moderate	Moderate
Pneumatic	Good	Moderate	Moderate
Vibration	Good	Cheap	Small

Table 9.1. Gyroscope comparison.

A comparison of some types of gyroscope is presented in Table 9.1. As the explosive development of design and manufacturing technologies continues, the characteristics of each type of gyroscope in Table 9.1 will improve over the years, and other new technologies may mature to the point where they can be used to further expand this table.

9.7 Inertial Measurement Unit (IMU)

The IMU is a single unit in the electronics module which collects angular velocity and linear acceleration data which is sent to the main processor. The IMU housing actually contains two separate sensors. The first sensor is the accelerometer triad. It generates three analogue signals describing the accelerations along each of its axes produced by, and acting on the vehicle. Due to the thruster system and physical limitations, the most significant of these sensed accelerations is caused by gravity. The second sensor is the angular rate sensor triad. It also outputs three analogue signals. These signals describe the vehicle angular rate about each of the sensor axes. Even though the IMU is not located at the vehicle centre of mass, the angular rate measurements are not affected by linear or angular accelerations. The data from these sensors is collected by the IMU microprocessor. The accelerometer triad and angular rate sensors within the IMU are mounted such that their sensor coordinate axes are not aligned with those of the vehicle. This is due to the fact that the two sensors in the IMU are mounted in two different orientations in the housing, along with the fact that the axes of the IMU are not aligned with the vehicle axes. We can replace the accelerometer triad by three accelerometers, and also the angular rate sensor triad by three gyrometers.

Inertial measurement components, which sense either acceleration or angular rate, are being embedded into common user interface devices more frequently as their cost continues to drop dramatically. These devices hold a number of advantages over other sensing technologies; they measure relevant parameters for human interfaces and can easily be embedded into wireless, mobile platforms.

IMU devices hold a number of advantages over other sensing technologies such as vision systems and magnetic trackers; they are small and robust, and can be made wireless using a lightweight radio-frequency link. However, in most cases, these inertial systems are put together in a very ad hoc fashion, where a small number of sensors are placed on known fixed axes, and the data analysis relies heavily on a priori information or fixed constraints. This requires a large amount of custom hardware and software engineering to be done for each application, with little reuse possible.

Specific applications for the IMU include unmanned vehicles (UVs), aircraft, missiles, marine and land navigation, imaging and camera stabilization, vehicle dynamics testing and self-guided systems.

9.8 Magnetic Compasses

A magnetic compass measure the Earth's magnetic field. When used in a positioning system, a compass measures the orientation of an object to which the compass is attached. The orientation is measured with respect to magnetic north. Magnetic field sensors are also know as *magnetometers*. The intensity of a magnetic field can be measured by the magnetic flux density B. The units of measurement are the tesla (T), the gauss (G), and the gamma (γ), where $1 \text{ T} = 10^4 \text{ G} = 10^9 \ \gamma$.

The Earth's magnetic field has an average strength of 0.5 G. It can be represented by a dipole that fluctuates in both time and space. The best-fit dipole location is approximately 440 km off-centre and is inclined approximately 11 degrees to the Earth's rotational axis [49]. The difference between true north and magnetic north is know as declination. This difference varies with both time and geographical location. To correct the difference, declination tables are provided on maps or charts for any given locale. For a vehicle location and navigation system, this does not cause a problem as long as all directions used in the system are referred to the same (magnetic or geographic) north [190].

Because of portability, high vibration durability, and quick response, electronic compasses have many advantages over conventional compasses. Among the electronic compasses, the fluxgate compass is the most popular one. The sensitivity range of this sensor is from 10^{-6} G to 100 G and the maximum frequency is about 10 kHZ. The term fluxgate is obtained from the gating action imposed by an ac-driven excitation coil that induces a time-varying permeability[6] in the sensor core.

Determining the magnetic field vector of the vehicle is simple. As the vehicle moves and changes its orientation relative to the Earth's magnetic field vector, the vehicle's own magnetic field vector moves with the vehicle. That means that regardless of vehicle orientation, the compass will see the sum of the vehicle vector (which remains constant) and a changing Earth vector. If the vehicle is moved through a complete 360 degrees closed-loop, the compass output will trace a circle. The coordinates of the centre of the circle will describe the vehicle magnetic field vector.

Fluxgate compass measurements generally contain two types of errors: short-term magnetic anomalies and long-term magnetic anomalies. Derivations caused by nearby power lines, big trucks, steel structures (such as freeway underpasses and tunnels), reinforced concrete buildings, and bridges

---

[6] Permeability is the property of a magnetizable substance that determines the degree to which it modifies the magnetic flux of the magnetic field in the region occupied by the magnetizable substance. This property in the magnetic field is analogous to conductivity in an electrical circuit.

are typical examples of short-term magnetic anomalies. Long-term magnetic anomalies result from inaccuracies in calibration, electrical/magnetical noise, or magnetization of the vehicle body.

9.9 Global Positioning System (GPS)

GPS is a weather-independent 24-hour position and navigation system that is maintained and operated by the U.S. Department of Defense (DoD) (Figure 9.15). The first satellites of this system were launched in 1978. The system achieved initial operability capability (IOC) in 1993 when the orbital constellation reached 24 space vehicles orbiting at an altitude of about 20180 km and 55 degrees inclined orbital planes [115, 129, 108, 85].

Fig. 9.15. The 24 satellites of the GPS.

DEFINITION

The Global Positioning System (GPS)[7] is a worldwide radio-navigation system formed from a constellation of 24 satellites and their ground stations. GPS is the only system today able to show you your exact position on the Earth anytime, in any weather, anywhere. GPS satellites orbit at 11,000 nautical miles above the Earth. They are continuously monitored by ground stations located worldwide. The satellites transmit signals that can be detected by anyone with a GPS receiver. Using the receiver, you can determine your location with great precision.

---

[7] GPS: a navigational system involving satellites and computers that can determine the latitude and longitude of a receiver on Earth by computing the time difference for signals from different satellites to reach the receiver [39].

GPS receivers have been miniaturized to just a few integrated circuits and so are becoming very economical. And that makes the technology accessible to virtually everyone. These days GPS is finding its way into cars, boats, aircraft, planes, construction equipment, movie-making gear, farm machinery, even laptop computers.

GPS CAPABILITIES

Currently GPS transmits two carriers, L1 = 1575.42 MHz (wavelength $\lambda_1 \approx$ 19.0 cm) and L2 = 1227.6 MHz (wavelength $\lambda_2 \approx$ 24.4 cm). The carriers are modulated with a precision (P) code (L1 and L2), and a coarse acquisition code C/A (L1). The P(Y)-code has been encrypted (Anti Spoofing AS) and is henceforth referred to as Y-code. The chipping rates for the P-code and C/A-code are 10.23 MHz and 1.023 MHz respectively. The C/A-code is normally available on L1 only, but could be activated on L2 by the ground control. Upon completion of the modernization phase, the GPS satellites are expected to transmit the C/A-code on L2 and have a new (third) civil signal, called L5 at 1176.45 MHz. Of course, L1 also carries the navigation message modulated at 50 bps [186, 129].

GPS provides two levels of service, the *Standard Positioning Service* and the *Precise Positioning Service* [181].

The Standard Positioning Service (SPS) is a positioning and timing service which will be available to all GPS users on a continuous, worldwide basis with no direct charge. SPS will be provided on the GPS L1 frequency which contains a coarse acquisition (C/A) code and a navigation data message. SPS provides a predictable positioning accuracy of 100 metres (95%) horizontally and 156 metres (95%) vertically and time transfer accuracy to UTC within 340 nanoseconds (95%).

The Precise Positioning Service (PPS) is a highly accurate military positioning, velocity and timing service which will be available on a continuous, worldwide basis to users authorized by the U.S. P(Y) code capable military user equipment provides a predictable positioning accuracy of at least 22 metres (95%) horizontally and 27.7 metres vertically and time transfer accuracy to UTC within 200 nanoseconds (95%). PPS will be the data transmitted on the GPS L1 and L2 frequencies. PPS was designed primarily for U.S. military use. It will be denied to unauthorized users by the use of cryptography. PPS will be made available to U.S. military and U.S. Federal Government users. Limited, non-Federal Government, civil use of PPS, both domestic and foreign, will be considered upon request and authorized on a case-by-case basis, provided:

• It is in the U.S. national interest to do so.
• Specific GPS security requirements can be met by the applicant.
• A reasonable alternative to the use of PPS is not available.

In Table 9.2, we summarize the characteristics of the GPS.

Item	Characteristic
Satellites	24 satellites broadcast signals autonomously
Orbits	Six planes, at 55 degrees inclination, each orbital plane includes four satellites at 20,180 km altitude, with a 12 hr period
Carrier frequencies	L1: 1575.42 MHz
	L2: 1227.60 MHz
Digital signals	C/A code (coarse acquisition code): 1.023 MHz
	P code (precise code): 10.23 MHz
	Navigation message: 50 bps
Position accuracy	SPS: 100 m horizontal and 156 m vertical (95%)
	PPS: 22 m horizontal and 27.7 m vertical (95%)
Velocity accuracy	SPS: 0.5 — 2 m/s observed
	PPS: 0.2 m/s
Time accuracy	SPS: 340 ns (95%)
	PPS: 200 ns (95%)

Table 9.2. GPS characteristics.

9.9.1 Elements

The GPS consists of three major segments: *space*, *control* and *user* [129].

The *space* segment consists of 24 operational satellites in six orbital planes (four satellites in each plane). The satellites operate in circular 20,200 km (10,900 nm) orbits at an inclination angle of 55° and with a 12-hour period. The position is therefore the same at the same sidereal time each day, i.e. the satellites appear 4 minutes earlier each day.

The *control* segment consists of five monitor stations (Hawaii, Kwajalein, Ascension Island, Diego Garcia, Colorado Springs), three ground antennae (Ascension Island, Diego Garcia, Kwajalein), and a Master Control Station (MCS) located at Schriever AFB in Colorado. The monitor stations passively track all satellites in view, accumulating ranging data. This information is processed at the MCS to determine satellite orbits and to update each satellite's navigation message. Updated information is transmitted to each satellite via the ground antennae.

The *user* segment consists of antennae and receiver-processors that provide positioning, velocity, and precise timing to the user.

How does GPS work?

GPS satellites circle the Earth twice a day in a very precise orbit and transmit signal information to Earth. GPS receivers take this information and use triangulation to calculate the user's exact location. Essentially, the GPS receiver compares the time a signal was transmitted by a satellite with the time it was received. The time difference tells the GPS receiver how far away the satellite is. Now, with distance measurements from a few more satellites, the receiver can determine the user's position and display it on the unit's electronic map.

A GPS receiver must be locked on to the signal of at least three satellites to calculate a 2D position (latitude and longitude) and track movement. With four or more satellites in view, the receiver can determine the user's 3D position (latitude, longitude and altitude). Once the user's position has been determined, the GPS unit can calculate other information, such as speed, bearing, track, trip distance, distance to destination, sunrise and sunset time and more.

SOURCES OF GPS SIGNAL ERRORS

Factors that can degrade the GPS signal and thus affect accuracy include the following:

Ionosphere and troposphere delays : The satellite signal slows as it passes through the atmosphere. The GPS system uses a built-in model that calculates an average amount of delay to partially correct for this type of error.

Signal multipath : This occurs when the GPS signal is reflected off objects such as tall buildings or large rock surfaces before it reaches the receiver. This increases the travel time of the signal, thereby causing errors.

Receiver clock errors : A receiver's built-in clock is not as accurate as the atomic clocks onboard the GPS satellites. Therefore, it may have very slight timing errors.

Orbital errors : Also known as ephemeris errors, these are inaccuracies of the satellite's reported location.

Number of satellites visible : The more satellites a GPS receiver can "see", the better the accuracy. Buildings, terrain, electronic interference, or sometimes even dense foliage can block signal reception, causing position errors or possibly no position reading at all. GPS units typically will not work indoors, underwater or underground.

Satellite geometry/shading : This refers to the relative position of the satellites at any given time. Ideal satellite geometry exists when the satellites are located at wide angles relative to each other. Poor geometry results when the satellites are located in a line or in a tight grouping.

Intentional degradation of the satellite signal : Selective Availability (SA) is an intentional degradation of the signal once imposed by the U.S. Department of Defense. SA was intended to prevent military adversaries from using the highly accurate GPS signals. The government turned off SA in May 2000, which significantly improved the accuracy of civilian GPS receivers.

Differential GPS (DGPS)

Differential GPS techniques can improve GPS performance significantly. The technique involves two GPS receivers. One is a master receiver at a reference station with known (surveyed) coordinates and the other is a receiver at a reference location whose coordinates are to be determined. The calculated position solution for the master receiver can be compared with the known coordinates to generate a differential correction for each satellite. In one such commonly used technique, the correction information can then be transmitted to the remote receiver for calculating the corrected local position. This effectively reduces the position error for commercial systems to under 15 m even when SA is in effect [190].

Table 9.3 shows the typical DGPS characteristics.

Item	Real-time characteristic
Receiver separation	<50 km
Data link update rate	5–10 sec
Position accuracy	15 m (2–5 m for low-noise receivers)
Velocity accuracy	0.1 m/s
Time accuracy	100 ns

Table 9.3. DGPS characteristics.

DGPS technology assumes that the range errors for the remote receiver are the same as for the master receiver. It is easy to see that this assumption may not be true all the time, especially when the two receiver are far apart. An accuracy level of 5 m can commonly be achieved for receiver separation of up to 50 km when low-noise receivers are used.

Wide-area DGPS (WADPGS) is another cost-effective method of bringing differential techniques to many users. Recall that DGPS needs to have a receiver separation of less than 50 km to provide very good accuracy. In other words, the main disadvantage of DGPS is the high correlation between the reference-station-to-user distance and the positioning accuracy achieved. To cover a wide area, a substantial increase in the number of reference stations is required. In contrast, WADGPS requires very few reference stations.

In addition to the Global Positioning System developed by the United States, other satellite systems have been developed that could be used for location and navigation, such as GLONASS, which was developed by the former USSR. NAVSAT has been developed by the European Space Agency (ESA), LOCSTAR is being promoted by the French space organization (CNES), and STARFIX has been developed by John E. Chance and Associates of Louisiana for the Gulf of Mexico and continental United States.

9.10 Vision Sensors

Vision sensors are video cameras with integrated signal processing and imaging electronics. They are used in industrial inspection, quality control, design and manufacturing diagnostic applications and recently in aircraft applications. They often include interfaces for programming and data output, and a variety of measurement and inspection functions [8, 57].

When specifying vision sensors, it is important to determine whether a monochrome or colour sensor is needed. Monochrome vision sensors present the image in black and white, or grayscale. Colour sensing vision sensors are able to read the spectrum range using varying combinations of different discrete colours. One common technique is sensing the red, green, and blue components (RGB) and combining them to create a wide spectrum of colours. Multiple chip colour is available on some vision sensors. It is a method of capturing colour in which multiple chips are each dedicated to capturing part of the colour image, such as one colour, and the results are combined to generate the full colour image. They typically employ colour separation devices such as beamsplitters rather than having integral filters on the sensors.

Important specifications to consider when searching for vision sensors include number of images stored and maximum inspection rate. The number of images stored represents captured images that can be stored in on-board memory or non-volatile storage. The maximum inspection rate is the maximum number of parts or process steps that can be inspected or evaluated per unit time. This is usually given in units of inspection per second. Other important parameters include horizontal resolution, maximum frame rate, shutter speed, sensitivity, and signal-to-noise ratio.

Inspection functions include object detection, edge detection, image direction, alignment, object measurement, object position, bar or matrix code, optical character recognition (OCR), and colour mark or colour recognition.

Imaging technology used in vision sensors includes CCD, CMOS, tube, and film. Charge Coupled Devices (CCD) use a light-sensitive material on a silicon chip to detect electrons excited by incoming light. They also contain integrated microcircuitry required to transfer the detected signal along a row of discrete picture elements (or pixels) and thereby scan an image very rapidly. CMOS image sensors operate at lower voltages than CCDs, reducing power consumption for portable applications.

Analogue and digital processing functions can be integrated readily onto the CMOS chip, reducing system package size and overall cost. In a tube camera, the image is formed on a fluorescent screen. It is then read by an electron beam in a raster scan pattern and converted to a voltage proportional to the image light intensity. With film technology the image is exposed onto photosensitive film, which is then developed to be played or stored. The shutter, a manual door that admits light to the film, typically controls exposure.

Other parameters to consider when specifying vision sensors include performance features, physical features, lens mounting, shutter control, sensor specifications, dimensions, and operating environment parameters.

Video cameras

Video cameras are used in machine vision, quality monitoring, security, and remote monitoring for industrial and commercial operations. Video cameras can operate in monochrome or colour.

Important performance specifications to consider when searching for video cameras include horizontal resolution, maximum frame rate, shutter speed, sensitivity, and signal-to-noise ratio. Horizontal resolution is the maximum number of individual picture elements that can be distinguished in a single scanning line. It is most common to characterize horizontal video resolution corrected for the image aspect ratio, or specify the resolution in the largest circle than can fit in a rectangular image. Thus, for example a 640×480 image would be specified as 480 horizontal lines. Maximum frame rate is the number of frames that can be captured per unit time, typically frames per second. Shutter time is the time of exposure or light collection. Typically this may be set across a wide range. Sensitivity refers to minimum scene illumination for good image quality.

The standard unit for illuminance is lux, or lumens per square metre (1 m/m^2). Signal-to-noise ratio is defined as the peak-to-peak camera signal output current to the RMS noise in the output current. This ratio represents how prevalent the noise component of a signal, and thus the image uncertainty, is in the total signal. Noise sources include sensor "dark current", electromagnetic interference, and any other spurious non-image signal elements. Higher SNR numbers represent less image degradation from noise.

Analogue video formats used by video cameras include NTSC, PAL, SECAM, RS170, RS330, and CCIR. Digital output interfaces common to vision sensors include RS232, RS422, RS485, parallel interfaces, Ethernet, De-viceNet, ARCNET, PROFIBUS, CANbus, Foundation Fieldbus, IEEE 1394 (Firewire), Modem, SCSI, TTL, USB, and radio or wireless. Choices for bits or pixels include 8 bits, 10 bits, 12 bits, 14 bits, or 16 bits. Colour outputs are typically RGB, Y PbPr, Y/C (S-video), or composite.

Other parameters to consider when specifying video cameras include specialty applications, performance features, physical features, lens mounting, shutter control, sensor specifications, dimensions, and operating environment parameters.

9.11 Ideal Sensor

It is clear that no single sensor can provide completely accurate vehicle position information. Therefore, multisensor integration is required in order to provide the on-vehicle system with complementary, sometimes redundant information for its location and navigation task. Many fusion technologies have been developed to fuse the complementary information from different sources into one representation format. The information to be combined may come from multiple sensors during a single period of time or from a single sensor over an extended period of time. Integrated multisensor systems have the potential to provide high levels of accuracy and fault tolerance [190].

Multisensor integration and fusion provide a system with additional benefits. These may include robust operational performance, extended spatial coverage, extended temporal coverage, an increased degree of confidence, improved detection performance, enhanced spatial resolution, improved reliability of systems operation, increased dimensionality, full utilization of resources, and reduced ambiguity.

9.12 Modems

A modem (short for **mo**dulator-**dem**odulator) is a device or program that enables a computer to transmit data over, for example, telephone or cable lines. Computer information is stored digitally, whereas information transmitted over telephone lines is transmitted in the form of analogue waves. A modem converts between these two forms. Fortunately, there is one standard interface for connecting external modems to computers, called RS-232 [75].

Consequently, any external modem can be attached to any computer that has an RS-232 port, which almost all personal computers have. There are also modems that come as an expansion board that you can insert into a vacant expansion slot. These are sometimes called onboard or internal modems.

While the modem interfaces are standardized, a number of different protocols for formatting data to be transmitted over telephone lines exist. Some, like CCITT V.34, are official standards, while others have been developed by private companies. Most modems have built-in support for the more common protocols at slow data transmission speeds at least, so most modems can communicate with each other. At high transmission speeds, however, the protocols are less standardized.

Aside from the transmission protocols that they support, the following characteristics distinguish one modem from another:

BPS: How fast the modem can transmit and receive data. At slow rates, modems are measured in terms of baud[8] rates. The slowest rate is 300 baud (about 25 cps). At higher speeds, modems are measured in terms of bits per second (bps). The fastest modems run at 115,200 bps, although they can achieve even higher data transfer rates by compressing the data. Obviously, the faster the transmission rate, the faster you can send and receive data. Note, however, that you cannot receive data any faster than it is being sent. If, for example, the device sending data to your computer is sending it at 2,400 bps, you must receive it at 2,400 bps. It does not always pay, therefore, to have a very fast modem. In addition, some telephone lines are unable to transmit data reliably at very high rates.

VOICE/DATA: Many modems support a switch to change between voice and data modes. In data mode, the modem acts like a regular modem. In voice mode, the modem acts like a regular telephone. Modems that support a voice/data switch have a built-in loudspeaker and microphone for voice communication.

---

[8] Baud: pronounced bawd, the number of signalling elements that occur each second. The term is named after J.M.E. Baudot, the inventor of the Baudot telegraph code.

At slow speeds, only one bit of information (signalling element) is encoded in each electrical change. The baud, therefore, indicates the number of bits per second that are transmitted. For example, 300 baud means that 300 bits are transmitted each second (abbreviated 300 bps). Assuming asynchronous communication, which requires 10 bits per character, this translates to 30 characters per second (cps). For slow rates (below 1,200 baud), you can divide the baud by 10 to see how many characters per second are sent.

At higher speeds, it is possible to encode more than one bit in each electrical change. 4,800 baud may allow 9,600 bits to be sent each second. At high data transfer speeds, therefore, data transmission rates are usually expressed in bits per second (bps) rather than baud. For example, a 9,600 bps modem may operate at only 2,400 baud.

AUTO-ANSWER : An auto-answer modem enables your computer to receive calls in your absence. This is only necessary if you are offering some type of computer service that people can call in to use.

DATA COMPRESSION : Some modems perform data compression, which enables them to send data at faster rates. However, the modem at the receiving end must be able to decompress the data using the same compression technique.

FLASH MEMORY : Some modems come with flash memory rather than conventional ROM, which means that the communications protocols can be easily updated if necessary.

FAX CAPABILITY: Most modern modems are fax modems, which means that they can send and receive faxes. To get the most out of a modem, you should have a communications software package, a program that simplifies the task of transferring data.

9.12.1 Radio Modems

Radio modems are radio frequency transceivers for serial data. They connect to serial ports (RS232, RS422, etc.) and transmit to and receive signals from other matching radio modems. Radio modems can be configured for internal or external mounting.

Internal radio modems are computer cards that are attached to the wired network through a computer that is part of the network. External radio modems are modules that are connected to the wired network by means of a physical port or interface.

Bus or interface types available for radio modems include Type II card, Type III card, CardBus, ISA, PCI, MIC, RJ-45, SC, ISDN BRI S/T interface, ISDN BRI U interface, serial ports (RS232, RS422, RS485), ST, USB, and PLC slot mount. The data rate is also important to consider. The data rate is the maximum data transfer rate at which the modem can deliver data. It is normally expressed in bits/second [57].

Important radio link specifications to consider when searching for radio modems include frequency band, operating mode, and radio technique.

Frequency band choices include 900 MHz, 2.4 GHz, 5 GHz, 23 GHz, VHF, and UHF.

Operating modes for radio modems include point-to-point, point-to-multi-point, and repeater mode. Point-to-point radio modems can transmit to only one modem/radio modem at a time. Point-to-multi-point modems can transmit to several modems/radio modems at a time.

Radio techniques include direct sequence spread spectrum and frequency hopping spread spectrum. Spread spectrum is a technique that is used to reduce the impact of localized frequency interferences. To achieve this, it uses more bandwidth than the system needs. There are two main spread spectrum modalities: direct sequence and frequency hopping. The principle of direct sequence spreads the signal on a larger band by multiplexing it with a code (signature) to minimize localized interference and noise. The system works over a large band. To spread the signal, each bit is modulated by a code. Frequency hopping uses a technique where the signal walks through a set of narrow channels in sequence.

The transmission frequency band is divided into a certain number of channels, and periodically the system hops to a new channel, following a predetermined cyclic hopping pattern. The system avoids interference by never staying in the same channel for a long period of time.

Common performance specifications for radio modems include full duplex transmission, maximum output power, number of channels, and sensitivity. Full duplex radio modems can transmit and receive at the same time. Maximum output power is the transmission power of the device. It is defined as the strength of the signals emitted, often measured in mW.

The number of channels defines the number of transmitting and receiving channels of the device. The sensitivity is the measure of the weakest signal that may be reliably sensed by the receiver. Sensitivity is measured in dBm, and the lower the value (higher in absolute value) the better is the receiver. Common features include antennae and RF connectors.

9.13 Microcontrollers

A microcontroller is a microprocessor[9] optimized to be used to control an electronic equipment or an embedded system.[10] Microcontrollers represent the vast majority of all computer chips sold; over 50% are simple controllers, and another 20% are more specialized DSPs (Digital Signal Processors).[11] They can be found in almost any electrical device, washing machines, microwave ovens, telephones, etc. [69, 133, 185].

---

[9] A microprocessor (abbreviated as μP or uP) is an electronic computer central processing unit (CPU) made from miniaturized transistors and other circuit elements on a single semiconductor integrated circuit (IC) (aka microchip or just chip).

[10] An embedded system is a special-purpose computer system built into a larger device. An embedded system is required to meet very different requirements than a general-purpose personal computer.

[11] A Digital Signal Processor (DSP) is a specialized microprocessor designed specifically for digital signal processing, generally in real time.

A microcontroller includes CPU, memory for the program (ROM), memory for data (RAM), I/O lines to communicate with peripherals and complementary resources, all in a closed chip. A microcontroller differs from a stand-alone CPU, because the first one generally is quite easy to make into a working computer, with a minimum of external support chips. The idea is that the microcontroller will be placed in the device to control, hooked up to power and any information it needs.

A traditional microprocessor won't allow you to do this. It expects all of these tasks to be handled by other chips. Some modern microcontrollers include a built-in high-level programming language; BASIC is quite common for this.

Fig. 9.16. Microcontroller ST62 of STMicroelectronics.

Microcontrollers (Figure 9.16) trade speed and flexibility for ease of use. There's only so much room on the chip to include functionality, so for every I/O device or memory the microcontroller includes, some other circuitry has to be removed. Finally, it must be mentioned that some microcontroller architectures are available from many different vendors in so many varieties that they could rightly belong to a category of their own.

Tables 9.4 and 9.5 show the common microcontrollers.

9.13.1 Fabrication Techniques

COMPLEMENTARY METAL OXIDE SEMICONDUCTOR – CMOS

This is the name of a common technique used to fabricate most of the newer microcontrollers. CMOS requires much less power than older fabrication techniques, which permits battery operation. CMOS chips also can be fully or near fully static, which means that the clock can be slowed up (or even stopped) putting the chip in sleep mode. CMOS has a much higher immunity to noise (power fluctuations or spikes) than the older fabrication techniques [65].

Microcontroller	Architecture
Amtel	ARM AVR
Renesas	H8
Holtek	HT8
Intel	8-bit $\left\{ \begin{array}{l} 8XC42 \\ MCS48 \\ MCS51 \\ 8xC251 \end{array} \right.$ 16-bit $\left\{ \begin{array}{l} MCS96 \\ MXS296 \end{array} \right.$ 32-bit $\left\{ i960 \right.$
National Semiconductor	COP8
Microchip	12-bit instruction PIC 14-bit instruction PIC $\left\{ PIC \ 16F84 \right.$ 16-bit instruction PIC

Table 9.4. Microcontrollers.

POST METAL PROGRAMMING – PMP – (NATIONAL SEMICONDUCTOR)

PMP is a high-energy implantation process that allows microcontroller ROM to be programmed after final metalization. Usually ROM is implemented in the second layer die, with nine or ten other layers then added on top. That means the ROM pattern must be specified early in the production process, and completed prototypes devices won't be available typically for six to eight weeks. With PMP, however, dies can be fully manufactured through metalization and electrical tests (only the passivation layers need to be added), and held in inventory. This means that ROM can be programmed late in the production cycle, making prototypes available in only two weeks [65].

9.13.2 Applications

Microcontrollers are frequently found in: appliances (microwave oven, refrigerators, television and VCRs, stereos), computers and computer equipment

Microcontroller	Architecture
Motorola	8-bit $\begin{cases} \text{68HC05 (CPU05)} \\ \text{68HC08 (CPU08)} \\ \text{68HC11 (CPU11)} \end{cases}$ 16-bit $\begin{cases} \text{68HC12 (CPU12)} \\ \text{68HC16 (CPU16)} \\ \text{Motorola DSP56800 (DSPcontroller)} \end{cases}$ 32-bit $\begin{cases} \text{Motorola 683XX (CPU32)} \\ \text{MPC500} \\ \text{MPC 860 (PowerQUICC)} \\ \text{MPC 8240/82500 (PowerQUICC II)} \end{cases}$
NEC	78K
Philips Semiconductors	LPC2000 LPC900 LPC700
STMicroelectronics	ST 62 ST 7
Texas Instruments	TSM370 MSP430
Western Design Center	8-bit $\{$ W65C02 – based μ Cs 16-bit $\{$ W65816 – based μ Cs
ZiLOG	Z8 Z86E02

Table 9.5. Microcontrollers.

(laser printers, modems, disk drives), automobiles (engine control, diagnostics, climate control), environmental control (greenhouse, factory, home), instrumentation, aerospace, and thousands of other uses. In many items, more than one processor can be found.

Microcontrollers are used extensively in robotics. In this application, many specific tasks might be distributed among a large number of controllers in one system. Communications between each controller and a central, possibly more powerful controller (or micro/mini/mainframe) would enable information to be processed by the central computer, or to be passed around to other controllers in the system.

A special application that microcontrollers are well suited for is data logging. Stick one of these chips out in the middle of a corn field or up in a ballon, and monitor and record environmental parameters (temperature, humidity, rain, etc). Small size, low power consumption, and flexibility make these devices ideal for unattended data monitoring and recording.

In order to get started with microcontrollers, several factors need to be considered. The factors are:

- cost
- convenience
- availability of development tools
- intended use.

Many manufacturers offer assembled evaluation kits or boards which usually allow you to use a PC as a host development system. Among some of the more popular evaluation kits/boards are:

- Parallax Basic Stamp
- Motorola EVBU, EVB, EVM, EVS
- Motorola 68750 starter kit
- Dallas Semiconductor DS5000TK
- Philips/CEIBO DS750
- American Educational Systems AES-51, AES-11, AES-88.

There exist free software for development, but you often get what you pay for. What is sorely lacking in freeware is technical support. Several packages are available that provide complete development environments for some of the more popular microcontrollers. If you want to be productive right away, think about investing $100. C isn't the only development system available, good solid Basic and Forth development systems are also available.

9.13.3 Microcontroller Programming Languages

Machine/Assembly language

Machine language is the program representation as the microcontroller understands it. *Assembly language* is a human-readable form of machine language which makes it much easier for us flesh and bone types to deal with. Each assembly language statement corresponds to one machine language statement (not counting macros).

An assembly/machine language program is fast and small. This is because you are in complete charge of what goes into the program.

Interpreters

An *interpreter* is a high-level language translator that is closer to natural language. The interpreter itself is a program that sits resident in the microcontroller. It executes a program by reading each language statement one at a time and then doing what the statement says to do. The two most popular interpreters for microcontrollers are BASIC and FORTH.

Compilers

A *compiler* is a high-level language translator that combines the programming ease of an interpreter with greater speed. This is accomplished by translating the program (on a host machine such as a desktop PC) directly into machine language. The machine language program is then burned onto an EPROM[12] or downloaded directly to the microcontroller. The microcontroller then executes the translated program directly, without having to interpret it first.

The most popular microcontroller compilers are C and BASIC. PL/M, from Intel, also has some popular support due to that company's extensive use of that language. Modula-2 has a loyal following due to its efficient code and high development productivity. Ada has many adherents among those designing on the larger chips (16 bits and above).

9.14 Real-time Operating System

A *Real-Time Operating System* or *RTOS* is an operating system that has been developed for real-time applications [62, 79, 84, 185]. Typically used for embedded applications they usually have the following characteristics:

- Small footprint (doesn't use much memory)
- Pre-emptable (any hardware event can cause a task to run)
- Multi-architecture (code ports to another type of CPU)
- Many have predictable response-times to electronic events.

Many real-time operating systems have scheduler and hardware driver designs that minimize the periods for which interrupts are disabled, a number sometimes called the interrupt latency.

Many also include special forms of memory management that limit the possibility of memory fragmentation, and assure a minimal upper bound on memory allocation and deallocation times.

It is a fallacy to believe that this type of operating system is "efficient" in the sense of having high throughput. The specialized scheduling algorithm and a high clock-interrupt rate can both interfere with throughput.

---

[12] EPROM – Erasable Programmable Read Only Memory.

9.14.1 Some Definitions

OPERATING SYSTEM

In computing, an operating system (OS) is the system software responsible for the direct control and management of hardware and basic system operations, as well as running applications such as word-processing programs and Web browsers [185].

REAL-TIME

There are several definitions of real-time, most of them contradictory. Unfortunately the topic is controversial, and there doesn't seem to be 100% agreement over the terminology.

POSIX[13] Standard 1003.1 defines real-time for operating systems as: the ability of the operating system to provide a required level of service in a bounded response time.

Another definition is:

An operation within a larger dynamic system is called a real-time operation if the combined reaction- and operation-time of a task is shorter than the maximum delay that is allowed, in view of circumstances outside the operation. The task must also occur before the system to be controlled becomes unstable. A real-time operation is not necessarily fast, as slow systems can allow slow real-time operations. This applies for all types of dynamically changing systems. The opposite of a real-time operation is a batch job [185].

PROGRAMMING LANGUAGE TO DEVELOP A REAL-TIME SYSTEM

The Ada community will always try to convince you their language is the best to use in any cases. Others will try to convince you to use an object-oriented language. The best solution is to avoid the use of dynamic object creation. Just create them at startup. The most used languages are (in alphabetical order): Ada, C, C++ for real-time system development. Most of the time small parts of the system are still written in assembler (small parts of device driver) [174].

The following list shows the most common real-time operating systems and embedded kernels [145].

- Allegro
- CMX

---

[13] Portable Operating System Interface (POSIX): a set of IEEE standards designed to provide application portability between Unix variants. IEEE 1003.1 defines a Unix-like operating system interface, IEEE 1003.2 defines the shell and utilities and IEEE 1003.4 defines real-time extensions.

- CRTX
- Harmony (National Research Council of Canada)
- PowerHawk
- Helios (Transputer, C40, ARM)
- INtime for Windows NT by Radisys
- iRMX by Radisys
- ITRON (consumer OS standard)
- Lynx OS (newsgroup)
- MAXION/OS by CCC
- Modcomp's UNIX and proprietary RT computers
- MTOS family of real-time operating systems (from IPI)
- Nucleus by OS/9 and OS-9000: Users Group in Chicago, European Users Group
- OSE family of operating systems
- Precise/MQX pSOS+
- QNX-RTOS QNX/Neutrino real-time POSIX kernel
- RTEMS (Real-Time Executive for Military Systems)
- RTMX
- RTX
- RTXDOS-16 (for embedded PCs) RTXDOS-32 (WIN32 API) TERSE (flow based, free and compact, for single and distributed microcontrollers)
- TXS-32 Real-time Extension (Windows NT for real-time and embedded computing)
- Vivaldi VRTX/OS 3.0
- VxWorks (newsgroup).

The most common companies selling real-time software are [145].

- Circuit Cellar
- CMX
- Dunfield development systems
- Forth, Inc. (makers of EXPRESS, polyFORTH, etc.)
- GenSym (maker of the G2 real-time expert system)
- Hi-Tech
- Integrated Systems (makers of pSOS+)
- Microtec Research
- National Instruments (makers of LabView)
- ObjectTime (maker of ROOM)
- Project Technology (Shlaer-Mellor OO software development)
- QNX Software Systems, Ltd
- Quiotix
- Real-Time Innovations (selling ControlShell)
- Technosoftware AG, Switzerland
- Verilog's ObjectGEODE
- VenturCom, selling Real-time Extensions

- Virtual-Time Software
- Whyron (emulators, compilers, debuggers)
- WindRiver (e.g. VxWorks)
- XLNT Designs.

A

Model Coefficients

Definition of the coefficients used in the helicopter dynamical model [9], equations (6.2)–(6.5) when mounted in a vertical platform:

$$c_0 = m$$

$$c_1 = I_{zzF} + I_{zzM} + 2m_T(x_T^2 + y_T^2)$$

$$c_2 = I_{zzM}$$

$$c_3 = I_{zzM} + r_\gamma^2 I_{yyT}$$

$$c_4 = -mg$$

$$c_5 = \frac{\rho p_M c_M a R_M^3}{6bl_M}$$

$$c_6 = \frac{\rho p_M c_M a R_M^2}{4} v_h$$

$$c_7 = 0.05\, mg$$

$$c_8 = \frac{\rho p_T c_T a R_T^3 r_\gamma^2 x_T}{6bl_T}$$

$$c_9 = \frac{\rho p_M c_M a R_M^3}{6bl_M} v_h$$

$$c_{10} = K_{engine}$$

$$c_{11} = \frac{\rho p_M c_M c_d R_M^4}{8}$$

$$c_{12} - \frac{5\rho p_M c_M a R_M^2}{16} v_h^2$$

where

$$a = \text{the lift-curve slope per degree of the wing}$$

$$b = \text{tail rotor blade length}$$

$$c_d = \text{drag coefficient}$$

$$c_M = \text{main rotor blade chord}$$
$$c_T = \text{tail rotor blade chord}$$
$$c_{\omega_M} = \text{damping coefficient due to the main rotor inflow}$$
$$c_{\omega_T} = \text{coefficient corresponding to the tail rotor inflow}$$
$$I_{yyT} = \text{inertia element of the tail rotor, with respect to } y\text{-axis}$$
$$I_{zzJ} = \text{inertia element of J} = \text{F (fuselage) or M (main rotor)},$$
$$\text{with respect to } z\text{-axis}$$
$$K_{engine} = \text{engine gain}$$
$$m = \text{helicopter total mass}$$
$$m_M = \text{main rotor blade mass}$$
$$m_T = \text{tail rotor blade mass}$$
$$r_\gamma = \text{gear ratio between the main and the tail rotors}$$
$$\rho = \text{air density}$$
$$R_M = \text{main rotor radius}$$
$$R_T = \text{tail rotor radius}$$
$$p_M = \text{number of blades of the main rotor}$$
$$p_T = \text{number of blades of the tail rotor}$$
$$v_h = \text{hover induced velocity}$$
$$x_T = X\text{-distance between the tail rotor centre and the reference}$$
$$\text{system fixed in the helicopter centre of mass}$$
$$y_T = Y\text{-distance between the tail rotor centre and the reference}$$
$$\text{system fixed in the helicopter centre of mass}$$

References

1. Aerosonde Robotic Aircraft. [Online] September 2004. Available at: http://www.aerosonde.com/index.php

2. AirScooter Corporation. AirScoot helicopter. [Online] September 2004. Available at: http://www.airscoot.com/pages/album.htm

3. Aldea M. and González M. *"MaRTE OS: An Ada kernel for real-time embedded applications"*. Proceedings of the International Conference on Reliable Software Technologies, Ada-Europe-2001, Leuven, Belgium, Lecture Notes in Computer Science, LNCS 2043, May 2001.

4. Alderete T. S., *"Simulator aero model implementation"*, [Online], NASA Ames Research Center, Moffett Field, California. Available at http://www.simlabs.arc.nasa.gov/library_docs/rt_sim_docs/Toms.pdf

5. Altuğ E., Ostrowski J. P. and Mahony R., *"Control of a quadrotor helicopter using visual feedback"*, Proceedings of the 2002 IEEE International Conference on Robotics and Automation, ICRA 2002, May 11–15, 2002, Washington, DC.

6. Analog Devices, *"Low cost ±2g/±10g dual axis iMEMS accelerometers with digital output"*, July 2004. Available at: http://www.analog.com/

7. Aström K. J. and Wittenmark B., **Computer-Controlled Systems: Theory and Design**, Prentice-Hall, Information and Systems Sciences Series, third edition, 1996. ISBN 0133148998.

8. Automated Imagining Association, *Machine vision online*. May 2004. http://www.machinevisiononline.org/

9. Avila-Vilchis J. C., *Modélisation et commande d'hélicoptère*, PhD Thesis, Institut National Polytechnique de Grenoble, Grenoble, France, 2001.

10. Avila J. C., Brogliato B., Dzul A. and Lozano R., *"Nonlinear modelling and control of helicopters"*, Automatica, 39(9):1583–1596, 2003.

11. Azouz N., Bestaoui Y. and Lematre O., *"Dynamic analysis of airships with small deformations"*, 3rd IEEE Workshop on Robot Motion and Control, Bukowy-Dworek, Nov. 2002, pp. 209–215.

12. Barnes W. McCormick, **Aerodynamics Aeronautics and Flight Mechanics**, John Wiley & Sons, New York, 1995. ISBN 0-471-57506-2.

13. Beji L., Abichou A. and Bestaoui Y., *"Position and attitude control of an underactuated airship"*, International Journal of Differential Equations and Applications, Vol. 8, Num. 3, pp. 231–255, 2004.

14. Beji L., Abichou A. and Bestaoui Y., *"Stabilization of a nonlinear underactuated autonomous airship – A combined averaging and backstepping approach"*, 3rd IEEE Workshop on Robot Motion and Control, Bukowy-Dworek, Nov. 2002, pp. 223–229.

15. Bendotti P. and Morris J. C., *"Identification and Stabilization of a Model Helicopter in Hover"*, Proceedings of the American Control Conference ACC'95, June 1995, Seattle, WA.

16. Bestaoui Y. and Hamel T., *"Dynamic modelling of small autonomous blimps"*, Conference MMAR, Methods and Models in Automation and Robotics, Miedzyzdroje, Poland, 2000, pp. 579–584.

17. Bestaoui Y. and Hima S., *"Some insights in path planning of small autonomous blimps"*, Archives of control sciences, vol. 11, 2001, pp. 139–166.

18. Bestaoui Y., Hima S. and Sentouh C., *"Motion planning of a fully actuated unmanned air vehicle"*, AIAA Conference on Navigation, Guidance and Control, Austin, Texas, Aug. 2003.

19. Bogdanov A., Carlsson M., Harvey G., Hunt J., Kieburtz R., Van der Merwe R., and Wan E., *"State-dependent Riccati equation control of a small unmanned helicopter"*, Proceedings of the AIAA Guidance Navigation and Control Conference, Austin, Texas, August 2003.

20. Brethé D. and Loiseau J.J. *"An effective algorithm for finite spectrum assignment of single-input systems with delays"*, Mathematics in computers and simulation, 45, 339–348, 1998.

21. Brian L. Stevens and Lewis F. L., **Aircraft Control and Simulation**, Wiley-Interscience Publication, ISBN 0-471-37145-9, 2nd edition, 2003.

22. Bridgman L., *Jane's all the world's aircraft 1954–55*. Jane's All The World's Aircraft Publishing Company, London, 1955.

23. Brockett R. W., *"Asymptotic stability and feedback stabilization"*, in Differential Geometric Control Theory, eds: R.W. Brockett, R. S. Millman, H.J. Sussmann, Birkhauser, Boston, 1983, pp. 181–191.

24. Castillo P., Lozano R., Fantoni I. and Dzul A., *"Control design for the PVTOL aircraft with arbitrary bounds on the acceleration"*, Proceedings of the IEEE 2002 Conference on Decision and Control CDC'02, December 2002, Las Vegas, Nevada.

25. Castillo P., Dzul A. and Lozano R., *"Real-time stabilization and tracking of a four rotor mini rotorcraft"*, IEEE Transactions on Control Systems Technology, Vol. 12, No. 4, pp. 510–516, July 2004.

26. Castillo P., *Modélisation et commande d'un hélicoptère à quatre rotors*, PhD Thesis, Université de Technologie de Compiègne, France, March 2004.

27. Cook M. V., Lipscombe J. M. and Goineau F., *"Analysis of the stability modes of the non rigid airship"*, Aeronautical Journal, 2000, pp. 279–289.

28. Coron J. M. and Rosier L., *"A relation between continuous time-varying and discontinuous feedback stabilization"*, J. Math. Estimations and Control, 1994, No. 4, pp. 67–84.

29. Coron J. M., *"On the stabilization of some nonlinear control systems: results, tools, and applications"*, NATO Advanced Study Institute, 1998, Montreal.

30. Curtis D. Johnson, **Process Control Instrumentation Technology**, Prentice Hall PTR, 2002. ISBN 0130602485.

31. Defense Update. International Online Defense Magazine. [Online] September 2004. Available at: http://www.defense-update.com/

32. Downing D. R. and Bryant W. H., *"Flight test of a digital controller used in a helicopter autoland system"*, Automatica, Vol. 23, No. 3, pp. 295–300, 1987.

33. Draganfly Innovations, [Online] September 2004. Available at: http://www.rctoys.com/

34. Dzul A., Hamel T. and Lozano R., *"Helicopter's nonlinear control via backstepping techniques"*, Proceedings of the ECC'01, Porto, Portugal, 2001.

35. Dzul A., Hamel T. and Lozano R., *"Modelling and nonlinear control for a coaxial helicopter"*, Proceedings of the IEEE 2002 International Conference on Systems, Man and Cybernetics, 6–9 October 2002, Hammamet, Tunisia.

36. Dzul A., *Commande automatique d'hélicoptères miniatures*, PhD Thesis, Université de Technologie de Compiègne, France, November 2002.

37. Dzul A., Lozano R. and Castillo P., *"Adaptive altitude control for a small helicopter in a vertical flying stand"*, International Journal of Adaptive Control and Signal Processing (IJACSP), Vol. 18, Issue 5, pp. 473–485, June 2004.

38. Elfes A., Siqueira Bueno S., Bergerman M. and Guinaraes Ramos J., *"A semiautonomous robotic airship for environmental monitoring mission"*, IEEE International Conference on Robotics and Automation, Detroit, MI, 1999, pp. 3449–3455.

39. Encyclopedia. The Free Dictionnary. [Online] May 2004. Available at: http://encyclopedia.thefreedictionary.com/

40. Enell Postcards. [Online] September 2004. Availalbe at: http://www.postcardpost.com/enel45.jpg

41. Etkin B. and Duff Reid L., **Dynamics of Flight**, John Wiley and Sons, New York, 1995. ISBN 0-471-03418-5.

42. Everett H. R., **Sensors for Mobile Robots: Theory and Application**, A K Peters, Wellesley, Massachusetts, 1995. ISBN 1-56881-048-2.

43. Fantoni I. and Lozano R., **Non-linear Control for Underactuated Mechanical Systems**, Communications and Control Engineering Series, Springer-Verlag, London, 2001. ISBN 1852334231.

44. Fantoni I. and Lozano R., *"Control of nonlinear mechanical systems"*, European Journal of Control, Vol 7: 328–348, 2001.

45. Fantoni I., Lozano R. and Castillo P., *"Stabilisation of the PVTOL aircraft"*, Proceedings of the 15th IFAC World Congress, July 2002, Barcelona, Spain.

46. Fastrack 3Space Polhemus, *User's Manual*, Colchester, Vermont, USA, 2001.

47. Fay J., **The Helicopter: History, Piloting, and How it Flies**. David Charles, London, 1976. ISBN 0715372491.

48. Foias C., Özbay H. and Tannenbaum A., **Robust Control of Infinite Dimensional Systems: Frequency Domain Methods**, Lecture Notes in Control and Information Sciences, No. 209, Springer-Verlag, London, 1996. ISBN 3-540-19994-2.

49. Fraden J., **AIP Handbook of Modern Sensors: Physics, Designs, and Applications**, New York, American Institute of Physics, 1993. ISBN 0387007504.

50. Frazzoli E., *Robust hybrid control for autonomous vehicle motion planning*, PhD Thesis, MIT, Cambridge, May, 2001.

51. Frazzoli E., Dahlen M. and Feron E., *"Trajectory tracking control design for autonomous helicopters using a backstepping algorithm"*, Proceedings of the American Control Conference ACC'00, pp. 4102–4107, June 28–30, 2000, Chicago, Illinois.

52. Fossen T., **Guidance and Control of Ocean Vehicle**, John Wiley and Sons, 1996. ISBN 0471941131.

53. Fuller A. T., *"In the large stability of relay and saturation control systems with linear controllers"*, Int. J. Control, 10(4):457–480, 1969.

54. Furutani E. and Araki M., *"Robust stability of state-predictive and Smith control systems for plants with a pure delay"*, International Journal of Robust and Nonlinear Control, Vol. 8, No. 18, pp. 907–919, 1998.

55. Gavrilets V., Martinos I., Mettler B. and Feron E., *"Control logic for automated aerobatic flight of a miniature helicopter"*, Proceedings of the AIAA Guidance, Navigation and Control Conference and Exhibit, AIAA GNC, Monterey, CA, August 2002.

56. Gavrilets V., Mettler B. and Feron E., *"Nonlinear model for a small size acrobatic helicopter"*, Proceedings of the AIAA Guidance, Navigation and Control Conference and Exhibit, AIAA 2001-4333, August 6–9, Montréal, Québec, Canada, 2001.

57. Global Spec, The engineering search engine. 2004. Available at: http://search.globalspec.com/Search/WebSearch

58. Gyroscope, adapted from Lecture 14, MIT, 2003. [Online] Available at: http://ocw.mit.edu/OcwWeb/Aeronautics-and-Astronautics/16-61Aerospace-DynamicsSpring2003/CourseHome/

59. Goldstein, H., **Classical Mechanics**, 2nd edition, Addison-Wesley, USA, 1980. ASIN: 0201029189.

60. Goodwin G. C. and Sang Sing K., **Adaptive Filtering Prediction and Control**, Information and Systems Sciences Series, Prentice-Hall, New Jersey, 1984. ASIN: 013004069X.

61. Gordon Leishman J., **Principles of Helicopter Aerodynamics**, Cambridge University Press, 2000. ISBN 0-5216606-0-2.

62. Hang Lee Y. and Krishna C.M., **Readings in Real-time Systems**, IEEE Computer Society Press, 1993. ISBN 0-8186-2997-5.

63. Hauser J., Sastry S. and Meyer G., *"Nonlinear control design for slightly nonminimum phase systems: Application to V/STOL aircraft"*, Automatica, 28(4):665–679, 1992.

64. Helicopter History Site, *History of Helicopters*, June 2004, Available at: http://www.helis.com

65. Hersch Russ, *"Embedded processor and microcontroller primer and FAQ"*, July 2004. Available at: http://www.faqs.org/faqs/microcontroller-faq/primer/

66. *Hiller Aviation Museum* [Online], June 2004. Available at: http://www.hiller.org/

67. Hima S. and Bestaoui Y., *"Motion generation on trim trajectories for an autonomous underactuated airship"*, 4th International Airship Conference, Cambridge, England, July 2002.

68. Hima S. and Bestaoui Y., *"Time optimal paths for lateral navigation of an autonomous underactuated airship"*, AIAA Conference on Navigation, Guidance and Control, Austin, Tx, 2003.

69. Hintz K. and Tabak D., **Microcontrollers: Architecture, Implementation, and Programming**, McGraw-Hill, 1992. ISBN 0-07-028977-8.

70. Horn R. A. and Johnson C. R., **Matrix Analysis**, Cambridge University Press, 1990. ISBN 0521386322.

71. Huang Ch. Y., Celi R. and Shih I-Ch., *"Reconfigurable flight control systems for a tandem rotor helicopter"*, Journal of the American Helicopter Society, Vol. 44, No. 1, pp 50–62, January 1999.

72. Hughes T. P., **Elmer Sperry: Inventor and Engineer**, Johns Hopkins University Press, London, 1971. ISBN 0801811333.

73. Hutchison C., *The airframe and producibility redesign of the T-Wing unmanned aerial vehicle*, Undergraduate Thesis, University of Sydney, NSW, Australia, November 2000.

74. Internet, *All the World's Rotorcraft*. May 2004. Available at: http://avia.russian.ee/index.html

75. Internet, *Webopedia*, May 2004. Available at: http://www.webopedia.com/

76. Isidori A., Marconi L. and Serrani A., *"Robust nonlinear motion control of a helicopter"*, IEEE Transactions on Automatic Control, Vol. 48, No. 3, pp. 413–426, 2003.

77. Johansson R., **System Modeling and Identification**, Prentice Hall, Information and System Sciences Series, 1993. ISBN 0-13-482308-7.

78. Johnson W., **Helicopter Theory**, Princeton University Press, 1980. ASIN: 0691079714.

79. Joseph M., **Real-Time Systems**, Prentice Hall Professional Technical Reference, 1996. ISBN 0-13-455297-0.

80. Jun M., Roumeliotis S. I. and Sukhatme G. S., *"State estimation via sensor modeling for helicopter control using an indirect Kalman filter"*, Proceedings of IEEE/RSJ International Conference on Intelligent Robots and Systems, pp. 1346–1353, October, Kyongju, Korea, 1999.

81. Kaloust J., Ham C. and Qu Z., *"Nonlinear autopilot control design for a 2-DOF helicopter model"*, IEEE Proceedings in Control Theory and Applications, pp. 612–616, Vol. 144, Issue 6, November 1997.

82. Kamen E. W., Khargonekar P. P. and Tannenbaum A., *"Proper stable bezout factorizations and feedback control of linear time delay systems"*, Int. J. Control, Vol. 43, No. 3, 837–857, 1986.

83. Khoury G. A. and Gillet J. D., **Airship Technology**, Cambridge University Press, 1999. ASIN: 0521430747.

84. Kavi K. M., **Real-time Systems: Abstractions, Languages and Design Methodologies**, IEEE Computer Society Press, 1992. ISBN 0-8186-3152-X.

85. Kayton M. and Fried W. R., **Avionics Navigation Systems**, John Wiley & Sons, USA, 1969. ISBN: 471-46180-6.

86. Kenneth A. and Tapscott R., *Studies of the lateral-directional flying qualities of a tandem helicopter in forward flight*, National Advisory Committee for Aeronautics, Report NACA-TR-1207, NASA, January 1954.

87. Khargonekar P. P., Poola K. and Tannenbaum A., *"Robust control of linear time invariant plants by periodic compensation"*, IEEE Trans. on Automatic Control, vol. AC-30, pp. 1088–1096, 1985.

88. Koo T. J. and Sastry S., *"Output tracking control design of a helicopter model based on approximate linearization"*, Proceedings of the IEEE Conference on Decision and Control CDC'98, pp. 3635–3640, Tampa, Florida, December 16–18, 1998.

89. Krstic M., Kanellakopoulos I. and Kokotovic P., **Nonlinear and Adaptive Control Design**, John Wiley & Sons, 1995. ISBN 0471127329.

90. La Civita M., Messner W. and Kanade T., *"Modelling of small-scale helicopters with integrated first-principles and system-identification techniques"*, Proceedings of the 58th American Helicopter Society Annual Forum, June 11–13, Montréal, Québec, Canada, 2002.

91. Landau I. D., Lozano R. and M'Saad M., **Adaptive Control**, Springer-Verlag, Communications and Control Engineering Series, 1997. ISBN 354076187X.

92. Lin F., Zhang W. and Brandt R. D., *"Robust hovering control of a PVTOL aircraft"*, IEEE Transactions on Control Systems Technology, 7(3):343–351, 1999.

93. Loiseau J. J., Mori K., Van-Assche V. and Lafay J. F., *"Feedback realization of compensators for a class of time-delay systems"*, Proceedings of the 38th Conference on Decision and Control, Phoenix, Arizona, 1999.

94. Lozano R, Brogliato B., Egeland O. and Maschke B., **Dissipative Systems Analysis and Control: Theory and Applications**. Springer-Verlag, Communications and Control Engineering Series, London, 2000. ISBN 1-85233-285-9.

95. Lozano R., Castillo P., Garcia P. and Dzul A., *"Robust prediction-based control for unstable delay systems: Application to the control yaw of a mini helicopter"*, Automatica, Elsevier Science, Vol. 40, No. 4, pp. 603–612, April 2004.

96. Lozano R., Castillo P. and Dzul A., *"Global stabilization of the PVTOL: Real-time application to a mini aircraft"*, International Journal of Control, Vol. 77, No. 8, pp. 735–740, May 2004.

97. Mahony R., Hamel T. and Dzul-López A., *"Hover control via Lyapunov control for an autonomous model helicopter"*, Proceedings of the 38th Conference on Decision and Control CDC'99, pp. 3490–3495, December 7–10 1999, Phoenix, Arizona.

98. Manitius, A. Z. and Olbrot A. W., *"Finite spectrum assignment problem for systems with delays"*, IEEE Trans. Autom. Contr., Vol. AC-24, No. 4, 541–553, 1979.

99. Marconi L., Isidori A. and Serrani A., *"Autonomous vertical landing on an oscillating platform: an internal-model based approach"*, Automatica, 38:21–32, 2002.

100. Martin P., Devasia S. and Paden B., *"A different look at output tracking: control of a VTOL aircraft"*, Automatica, 32(1):101–107, 1996.

101. Mathie, M. J., Coster A. C. F., Lovell N. H. and Celler B. G., *Accelerometry: providing an integrated, practical method for long-term, ambulatory monitoring of human movement*, Institute of Physics Publishing, Physiological Measurement, February 2004.

102. Mazec F. and Praly L., *"Adding integrations, saturated controls, and stabilization for feedforward systems"*, IEEE Transactions on Automatic Control, 41(11):1559–1578, 1996.

103. Mesbahi M. and Hadaegh F. Y., *"A robust approach for the formation flying of multiple spacecraft"*, Proceedings of the European Control Conference ECC'99, Karlsruhe, Germany, 31 Aug – 3 Sep, 1999.

104. Mettler B., Tischler M. and Kanade T., *"System identification modelling of a model scale helicopter for the development of high performance helicopter based unmanned aerial vehicles"*, Journal of the American Helicopter Society, 47(1):50–63, 2002.

105. Mondié S., Dambrine M. and Santos O. "*Approximation of control laws with distributed delays: a necessary condition for stability*", IFAC Conference on Systems, Structure and Control, Prague, Czek Republic, 2001.

106. Mondié S., Lozano, R. and Collado J., "*Resetting process-model control for unstable systems with delay*", 40th IEEE Conference on Decision and Control, Orlando, Florida, 2001.

107. Mondié S., García P. and Lozano R., "*Resetting Smith predictor for the control of unstable systems with delay*", IFAC 15th Triennial World Congress, Barcelona, Spain, 2002.

108. Montenbruck O. and Gill E., **Satellite Orbits**, Springer-Verlag, Germany, 2001. ISBN 3-540-67280-X.

109. Morin P. and Samson C., "*Time varying exponential stabilization of the attitude of a rigid spacecraft with two controls*", IEEE Conf. on Decision and Control, New Orleans, LA, 1995, pp. 3988–3993.

110. Morris J., Van Nieuwstadt M. and Bendotti P., "*Identification and control of a model helicopter in hover*", Proceedings of the American Control Conference ACC'94, Baltimore, Maryland, 1994.

111. Morse A. S., "*Ring models for delay-differential systems*", Automatica, Vol. 12, 529–531, 1976.

112. Mullhaupt P., Srinivasan B., Lévine J. and Bonvin D., "*Cascade control of the toycopter*", European Control Conference ECC'99, 31 Aug – 3 Sep 1999, Karlsruhe, Germany.

113. Munson K., *Jane's unmanned aerial vehicles and targets*, Jane's Information Group, Sentinal House, 163 Brighton Road, Coulsdon, Surrey CR5 2NH, UK, 1998.

114. Murray R. M., **Control in an Information Rich World: Report of the Panel on Future Directions in Control, Dynamics, and Systems**, Softcover, June 2002. ISBN 0-89871-528-8.

115. National Academy Press, **Technology for Small Spacecraft**, Washington, DC, 1994. ISBN 0-309-05075-8.

116. National Air and Space Museum, *Centennial of flight*, July 2004. Available at: http://www.centennialofflight.gov/index.htm

117. National Highway Traffic Safety, *Technology alternatives for an automated collision notification system*, Report, U.S. Department of Transportation, Springfield, Virginia, August 1994.

118. National Instruments, *Accelerometer Principles*. [Online] September 2004. Available at: http://www.ni.com/

119. National Museum of Naval Aviation, [Online] November 2004. Available at: http://naval.aviation.museum/home.html

120. National Museum of Science and Technology, *Leonardo Da Vinci*, July 2004. Available at: http://www.museoscienza.org/english/leonardo/vite.html

121. National Museum of the United States Air Force. Available at http://www.wpafb.af.mil/museum/

122. Nelson R. C., **Flight Stability and Automatic Control**, second edition, McGraw-Hill Science/Engineering/Math, 1997. ISBN 0070462739.

123. Niculescu S., **Delay Effects on Stability: A Robust Control Approach**, Springer-Verlag, Heidelberg, Germany, 2001. ISBN 1852332913.

124. Niculescu S. and Lozano R., *"On the passivity of linear delay systems"*, IEEE Transactions on Automatic Control, Vol. 46, Issue 3, pp. 460–464, March 2001.

125. Olfati-Saber R., *"Global configuration stabilization for the VTOL aircraft with strong input coupling"*. Proceedings of the 39th IEEE Conf. on Decision and Control, Sydney, Australia, Dec. 1999.

126. Olfati-Saber R., *Nonlinear control of underactuated mechanical systems with application to robotics and aerospace vehicles*, PhD Thesis, MIT, Cambridge, Ma., February 2001.

127. Ousingsawat J. and Campbell M. E., *"On-line estimation and path planning for multiple vehicles in an uncertain environment"*, International Journal of Robust and Nonlinear Control, Vol. 14, pp. 741–766, 2004.

128. Padfield G. D., **Helicopter Flight Dynamics: The Theory and Application of Flying Qualities and Simulation Modeling**, American Institute of Aeronautics and Astronautics, 1996. ISBN 1563472058.

129. Parkinson B. W. and Spilker J. J. Jr., **Global Positioning System: Theory and Applications, Vol: I & II, Progress in astronautics and aeronautics**, Cambridge, Massachusetts, 1996. ISBN 1-56347-107-8.

130. Palmor Z. J., *"Time-delay compensation – Smith predictor and its modifications"*, in: S. Levine (Ed.), The Control Handbook, CRC Press, Boca Raton FL, 1996, pp. 224–237.

131. Palomino A., Castillo P., Fantoni I., Lozano R. and Pégard C., *"Control strategy using vision for the stabilization of an experimental PVTOL aircraft setup"*, CDC'03, Maui, Hawaii, December 2003.

132. Paiva E. C., Bueno S. S. and Gomes S. B. V., *"A control system development environment for AURORA's semi-autonomous robotic airship"*, IEEE Conference on Robotics and Automation, Detroit, MI, 1999, pp. 2328–2335.

133. Peatman J. B., **Design with Microcontrollers**, McGraw-Hill College, 1988, ISBN 0070492387.

134. Pettersen K. Y. and Egeland O., *"Time-varying exponential stabilization of the position and attitude of an underactuated autonomous underwater vehicle"*, IEEE Transactions on Automatic Control, 44(1):112–115, January 1999.

135. PCS Edventures, The educational adventure where kids discover how things work. [Online] September 2004. Available at: http://www.discover.edventures.com/

136. Pilotfriend, Century of flight. [Online] September 2004. Available at: http://www.pilotfriend.com/century-of-flight/index.htm

137. Pitt D. M. and Peters D. A., *"Theoretical prediction of dynamic inflow deriva-tives"*, Vertica, Vol. 5, No. 1 pp. 21–34, March 1981.

138. Postlethwaite I., Konstantopoulos I., Sun X-D., Walker D. and Alford A., *"Design, flight simulation, and handling qualities evaluation of an LPV gain-scheduled helicopter flight control system"*, European Control Conference ECC'99, 31 Aug – 3 Sep 1999, Karlsruhe, Germany.

139. Pounds P., Mahony R., Hynes P. and Roberts J., *"Design of a four rotor aerial robot"*, Proceedings of the Australasian Conference on Robotics and Automation, Auckland, Australia, 2002.

140. POSIX.13, IEEE Std. 1003.13-1998. *Information Technology – Standardized Application Enviroment Profile – POSIX Realtime Application Support (AEP)*. Institute of Electrical and Electronics Engineers, 1998.

141. Prouty R. W., **Helicopter Performance, Stability and Control**, Krieger, 1990. ISBN 1-57524-209-5.

142. Radix J.C., **Gyroscopes et gyromètres**, CEPADUES Editions, Ecolé National Superieur de l'Aéronautique et de l'espace, 1997.

143. *Real Time Workshop Manual*, The Math Works, 1998.

144. Reyhanoglu M., Van der Schaft A., McClamroch N. H. and Kolmanovsky I., *"Dynamics and control of a class of underactuated mechanical systems"*, IEEE Transactions on Automatic Control, Vol. 44, No. 09, p. 1663, September 1999.

145. Riegler A., *Embedded and Real-Time Systems*, Artificial Intelligence Labora-tory, Department of Information Technology, University of Zürich, Switzerland. July 2004. Available at:
http://www.ifi.unizh.ch/groups/ailab/links/embedded.html

146. Sadler G., *Early History of the Helicopter*, PhD Thesis, Australia. 1995.

147. Sepulchre R., Jankovic, and Kokotovic P., **Constructive Nonlinear Con-trol**, Springer-Verlag, London, 1997. ISBN 3540761276.

148. Shavrov V. B., *"History of aircraft construction in the USSR"*, Mashinostroe-nie, Moscow. ISBN 5-217-02528-X.

149. SIMULINK Manual. The Math Works, 1998.

150. Sira-Ramirez H., Zribi M. and Ahmad S., *"Dynamical sliding mode control approach for vertical flight regulation in helicopters"*, IEEE Control Theory and Applications, Vol. 141, No. 1, pp. 19–24, January 1994.

151. Sira-Ramirez H. and Castro-Linares R., *"On the regulation of a helicopter system: a trajectory planning approach for the Liouvillian model"*, European Control Conference ECC'99, 31 Aug – 3 Sep 1999, Karlsruhe, Germany.

152. Sikorsky Aircraft Corp., [Online] September 2004. Available at:
http://www.sikorsky.com/sac/Home/0,3170,CLI1_DIV69_ETI541,00.html

153. Slotine J.-J. and Li W., **Applied Nonlinear Control**, Prentice-Hall Inter-national Editions, 1990. ISBN 0130408905.

154. Smith O. J. M., *"Closer control of loops with dead time"*, Chem. Eng. Prog., Vol. 53, pp. 217–219, 1959.

155. Sridhar B. and Lindorff P., *"Application of pole-placement theory to helicopter stabilization systems"*, Proceedings of the 6th Hawaii International Conference on Systems Sciences, Western Periodicals, North Hollywood, CA, pp. 405–407, 1973.

156. Stengel R. F., Broussard J. R. and Berry P. W., *"Digital flight control design for a tandem-rotor helicopter"*, Automatica, Vol. 14, No. 4, pp. 301–312, July 1978.

157. Stepniewsky W. Z. and Keys C. N., **Rotary-Wing Aerodynamics**, Dover Publishing, New York, 1984. ISBN 0486646475.

158. Stone H. and Wong K. C. *"Preliminary design of a tandem-wing tail-sitter UAV using multi-disciplinary design optimisation"*, International Aerospace Congress, Sydney, pp. 707–720, February 1997.

159. Stone H., Coates E., Clarke G., Wong K. C. and Gibbens P. W. *"A vertical take-off and landing UAV for surveillance applications"*, International Aerospace Congress, Adelaide, September 1999.

160. Stone H., Configuration design of a canard configured tail-sitter unmanned vehicle using multidisciplinary optimisation, PhD Thesis, University of Sydney, Sydney, Australia, 1999.

161. Stone H. and Clarke G. *"The T-wing: a VTOL UAV for defense and civilian applications"*, UAV Australian Conference, Melbourne, February 2001.

162. Stone H. and Clarke G. *"Optimization of transition maneuvers for a tail-sitter unmanned air vehicle (UAV)"*, Australian International Aerospace Congress, Canberra, Australia, March 2001.

163. Stone H., *"Aerodynamic modeling and simulation of a wing-in-slipstream tail-sitter UAV"*, Biennial AIAA International Powered Lift Conference, Williamsburg, Virginia, 2–4 November 2002.

164. Stone H., *"The T-wing tail-sitter research UAV"*, Biennial AIAA International Powered Lift Conference, Williamsburg, Virginia, 2–4 November 2002.

165. Stone H., *"The T-wing tail-sitter UAV"*, Flight International UAV Australia Conference, Melbourne, February 2003.

166. Stone H., *"Modeling, simulation and control design for the T-wing tail-sitter UAV"*, 4th Australian Pacific Vertiflite Conference on Helicopter Technology, Melbourne, 21–23 July 2003.

167. Stone H., *"Design considerations for a wing-in-slipstream tail-sitter unmanned air vehicle (UAV)"*, 10th Australian International Aerospace Congress, Brisbane, 28 July – 1 August 2003.

168. Stone H., *"Transition maneuver optimization for the T-wing tail-sitter UAV"*, 10th Australian International Aerospace Congress, Brisbane, 28 July – 1 August 2003.

169. Tanaka K., Ohtake H. and Wang O. H., *"A practical design approach to stabilization of a 3-DOF RC helicopter"*, IEEE Transactions on Control Systems Technology, Vol. 12, No. 2, pp. 315–325, March 2004.

170. Tapscott R. and Kenneth A., *Studies of the speed stability of a tandem helicopter in forward flight*, National Advisory Committee for Aeronautics, Report NACA-TR-1260, NASA, January 1, 1956.

171. Tapscott R., *Some static longitudinal stability characteristics of an overlapped-type tandem-rotor helicopter at low airspeeds*, Technical Note 4393, National Advisory Committee for Aeronautics, NASA, September 1958, Washington, USA.

172. Teel A. R., *"Global stabilization and restricted tracking for multiple integrators with bounded controls"*. Systems and Control Letters, Vol. 18, pp. 165–171, 1992.

173. Teel A. R., *"A nonlinear small gain theorem for the analysis of control systems with saturation"*. IEEE Transactions on Automatic Control, Vol. 41, No. 9, pp. 1256–1270, 1996.

174. Timmerman M. et al., *Comp. realtime: Frequently Asked Questions (FAQs)*, Dedicated Systems Experts NV, Belgium, July 2004. Available at: http://www.faqs.org/faqs/realtime-computing/faq/

175. Tischler M. and Cauffman M., *"Frequency-response method for rotorcraft system identification: flight applications to BO-105 coupled rotor/fuselage dynamics"*, Journal of the American Helicopter Society, Vol. 37, No. 3, pp. 3–17, 1992.

176. Titterton D. H. and Weston J. L., **Strapdown inertial navigation technology**, Peter Peregrinus on behalf of the Institution of Electrical Engineers, London, United Kingdom, 1997, ISBN 0 86341 260 2.

177. Thomasson P., *"Equations of motion of a vehicle in a moving fluid"*, Journal of Aircraft, vol 37, No. 4, pp. 631–639, 2000.

178. Tsourdos A. and White B. A., *"Lateral acceleration control design for an LPV missile model"*, Proceedings of the European Control Conference ECC'99, Karlsruhe, Germany, 31 Aug – 3 Sep. 1999.

179. UAVs. *New world vistas: Air and space for the 21st centry*, Human Systems and Biotechnology Systems, Vol. 7, No. 0, pp. 17–18, 1997.

180. *UAV forum*, July 2004. Available at: http://www.uavforum.com/library/photo.htm

181. USNO. *NAVSTAR GPS Operations*. May 2004. Available at: http://tycho.usno.navy.mil/gpsinfo.html

182. Van-Assche V., *Étude et mise en oeuvre de commandes distribuées*, École Central de Nantes, PhD Thesis, Université de Nantes, France, 2002.

183. Watanabe K. and Ito M., *"A process model control for linear systems with delay"*, IEEE Trans. Autom. Contr., Vol. AC-26, No. 6, 1261–1268, 1981.

184. Weilenmann M. and Geering H., *"A test bench for rotorcraft hover control"*, Journal of Guidance, Control and Dynamics, Vol. 17, pp. 729–736, 1994.

185. Wikipedia. The free encyclopedia. [Online] May 2004. Available at: http://en.wikipedia.org/wiki/Main_Page

186. Witchayangkoon B., *Elements of GPS precise point positioning*, PhD Thesis, University of Maine, USA December 2000.

187. Yeates J. E., *Flight measurements of the vibration experienced by a tandem helicopter in transition, vortex-ring state, landing approach, and tawed flight*, Technical Note 4409, National Advisory Committee for Aeronautics, NASA, Washington, USA, September 1958.

188. Young L. A., Aiken E. W., Johnson J. L., Demblewski R., Andrews J. and Klem J., *"New concepts and perspectives on micro-rotorcraft and small autonomous rotary-wing vehicles"*, Proceedings of the 20th AIAA Applied Aerodynamics Conference, St. Louis, MO, 2002.

189. Zhao J. and Kanellakopoulos I., *"Active identification for discrete-time nonlinear control – Part I: output-feedback systems"*, IEEE Transactions on Automatic Control, Vol. 47, No. 2, pp. 210–224, 2002.

190. Zhao Yilin, **Vehicle Location and Navigation Systems**, Artech House, Boston, London, 1997. ISBN 0-89006-861-5.

Index